BIOTECHNOLOGY

MJP
PUBLISHERS

BIOTECHNOLOGY

Prakash S. Lohar

Reader, Department of Zoology
M.G.S. M's Arts Science & Commerce College
Chopda, Jalgaon

Chennai Trichy Tirunelveli NewDelhi

MJP Publishers

All rights reserved   No. 44, Nallathambi Street,

Printed and bound in India   Triplicane, Chennai 600 005

MJP 006   © Publishers, 2023

Publisher : C. Janarthanan

This book has been published in good faith that the work of the author is original. All efforts have been taken to make the material error-free. However, the author and publisher disclaim responsibility for any inadvertent errors.

To
Dr. P.S. Hardia

FOREWORD

Biotechnology in simple words means obtaining 'products' and 'services' from 'biological agents'. Microorganisms and isolated animal and plant cells, in majority of cases, act as biological agents. In fact, they have become tiny, self-contained factories faithfully churning out dictated products. Biotechnology has already made a significant impact in the areas of agriculture, medicine and pharmaceuticals, industrial chemicals, pollution control, etc. Transgenic crops resistant to either pests or pesticides, tissue cultured plants, gene therapy, production of industrial enzymes, mapping of human genome and cloning of animals and humans are some of the success stories in biotechnology today. All this is made possible, to a large extent, due to our ability to manipulate the genetic material besides our understanding of its structure and functional aspects.

Looking at the scope and applications of biotechnology, it has been incorporated in the curricula of undergraduate programmes of Botany, Zoology, Biochemistry, Microbiology, Pharmacy, etc. There was a need for a textbook which can provide a comprehensive view of biotechnology and its various applications to the students of these courses. The book, *Biotechnology* written by Dr. Prakash S. Lohar is therefore a welcome step in the right direction. The book has been divided in well-thought-of eleven chapters. Each chapter, in turn, has been presented in simple language giving illustrations wherever necessary.

After going through the contents of the book, I am sure that the book will be well-received by the undergraduate students of Zoology, Botany, Microbiology, Biotechnology, Biochemistry and Pharmacy. Also, the postgraduate students of these subjects may find it a useful source of information. I must congratulate Dr. Lohar for taking genuine efforts in bringing out such a useful and much needed book for the beginners as well as the advanced learners in the subject of Life Sciences in general and biotechnology in particular.

Dr. V. L. Maheshwari
Head of the Department
Department of Biochemistry
School of Life Sciences
North Maharashtra University
Jalgaon

PREFACE

Science affects our everyday lives in thousands of different ways and the scientific advances of today will shape our lives in the future. The fascinating discovery of the structure of DNA laid the foundation of modern bioscience including Biotechnology that is transforming our lives in one or many ways.

Biotechnology-based industries such as health care and pharmaceutical, crop and livestock breeding and use of microorganisms and plants to produce valuable new materials provide tremendous benefits to mankind either directly or indirectly. The momentous discovery of the three-dimensional structure of DNA made by Watson and Crick, ultimately paved the way for the Human Genome Project. The rapid progress in the field of genetics has profoundly changed our understanding of the basic processes of life. This exciting new knowledge is already pointing to gene level treatments in medicine, safer food, improved crops, healthier farm animals and a cleaner environment. Biotechnology affects over 30% of the global economic turnover by the way of health care, food and energy, agriculture and forestry, and this economic impact will grow as biotechnology provides new methods to process the raw materials. Biotechnology is developing at a phenomenal pace and will increasingly become a necessary part of modern life.

Modern biotechnology involves improved methods for production of antibiotics, vaccines, monoclonal antibodies, new molecular innovations due to genetic engineering, raising transgenic plants and animals, gene therapy in humans to eradicate many diseases, and biological fuel production. It also contains discussion on ethical, socio-economic, environmental, health and safety implications of fascinating and powerful applications of the latest scientific innovations in the field of biotechnology.

Prakash S. Lohar

ACKNOWLEDGEMENTS

My thanks are due to Dr. Maheshwari, Head, Department of Biochemistry, School of Life Sciences, North Maharashtra University, Jalgaon for providing the foreword for this book.

I am thankful to Hon'ble Dr. S. G. Patil, President, Mr. V.D. Joshi, Vice-President, Mr. S. S. Patil, Secretary, Dr. D. D. Patil, Principal, Mr. D. B. Deshmukh, Vice-Principal, Mr. S. S. Alizad, Vice-Principal, and Dr. P. B. Pawar, Former Principal of M. G. T. S. M's Arts, Science and Commerce College, Jalgaon for their continuous support and cooperation in this endeavour.

I would like to record special thanks to Mr. Rajesh Chavan (Scientist, Alembic Pharmaceutical), Dr. Swati Chavan, Mr. Mahesh Lohar and Mr. Bhushan Lohar from the fields of Medical Microbiology, Polymer Chemistry and Plant biotechnology respectively, who have helped me to incorporate latest and authentic information from their specialised areas. I am also thankful to Mr. S. R. Patil (PG student from Dept. of Zoology, Pune University) and Mr. S. N. Potdar (PG student from Dept. of Bioinformatics, Pune University) for their extended help to complete this book.

I express my deep sense of gratitude to Dr. B. D. Singh, Professor from school of Biotechnology, Faculty of Science, Banaras Hindu University, Varanasi, Prof. S. N. Jogdand, Dept. of Microbiology,

Modern College, Vashi, New Mumbai, Dr. K. K. De, Dept. of Botany, Barasat Govt. College, West Bengal from whom I got inspiration to complete this manuscript.

I extend my sincere thanks to Prof. A. S. Patil, Chairman of BOS, North Maharashtra University, the Staff of M. G. S. M's Arts Science and Commerce College, and to my friends and relatives for their consant encouragement to complete this work. I am also thankful to Rishikesh Bhavsar from Pankaj Computer Institute, Chopda for providing computer facility for typing and drawing figures in the manuscript.

I am especially grateful to my wife Savita, Daughter Shivani and son Kaushal who have supported me during this venture.

I express my heartiest thanks and indebtedness to Mr. J.C. Pillai, Director, Mr. C. Sajeeshkumar, Managing Editor and Ms. Parvath Radha, Project Coordinator of MJP Publishers, Chennai for their keen interest, encouragement and editorial efforts for publishing this book.

Dr. Prakash S. Lohar

CONTENTS

1

INTRODUCTION

1.1 DEFINITION

We are living in a fast-paced age. The increasing diversity of science and a set of techniques play an indispensable role in modern life. Like most disciplines, biotechnology holds high potential for solving many of the problems our world faces. The term 'biotechnology' is the combination of 'biology' and 'technology'. It deals with the exploitation of biological agents for generating useful products and services. In other words, it is a discipline which enables its exponents to convert raw materials to final products where either the raw material and/or a stage in the production process involves biological entities.

Some selected definitions of biotechnology are given below.

❑ The European Federation of Biotechnology (EFB) considers biotechnology as "the integrated use of biochemistry, microbiology and engineering sciences in order to achieve technological (industrial) application capabilities of microorganisms, cultured tissues/cells and parts thereof."

❑ U.S. National Science Foundation defines biotechnology as "the controlled use of biological agents, such as microorganisms or cellular components for beneficial use."

❑ The Spinks Reports (1980) defined biotechnology as "the application of biological organisms, systems or processes to manufacturing and service industries."

❑ According to Gibbs and Greenhalgh (1983) biotechnology may be defined as "the use of living organisms in systems or processes for the manufacture of useful products; it may involve algae, bacteria, fungi, yeast, cells of higher plants and animals or subsystems of any of these or isolated components from living matter."

❑ J.D.Bulock (1987) defined biotechnology as "the controlled and deliberate application of simple biological agents—living or dead cells or cell components—in technically useful operations either of productive manufacture or as service operation."

Thus biotechnology has been defined in many forms but each of them involves two common factors. First, the use of biological agents and second, the product or service is generated for the well-being of humans.

1.2 BIOTECHNOLOGY—AN INTERDISCIPLINARY ACTIVITY

Biotechnology is an interdisciplinary pursuit. In recent decades a characteristic feature of the development of science and technology has been the increasing resort to multidisciplinary strategies for the solution of various problems. The term 'multidisciplinary' deals with the quantitative extension of approach to the problems that commonly occur within a given area.

Biotechnology encompasses several disciplines of basic sciences and engineering. Biotechnologists on one hand can utilise the scientific discipline such as chemistry, microbiology, biochemistry, genetics, cell and molecular biology, immunology, physiology, electronics and computer science while on the other hand, they depend heavily on process, chemical, mechanical and biological engineering (figure 1.1), since the involvement of microorganisms and cells, and their processing (the manufacture of commercial products on large-scale, etc.) is based on them.

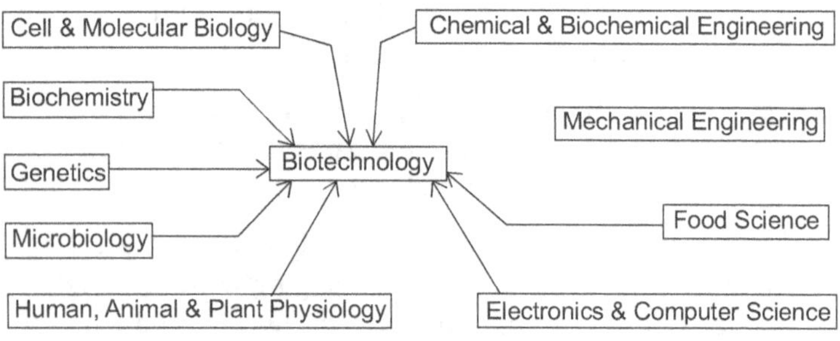

Figure 1.1   The interdisciplinary nature of biotechnology

1.3 BIOTECHNOLOGY—SCOPE AND IMPORTANCE

The scientific and technological discoveries have led human beings to the brink of a new industrial revolution—the bioindustrial revolution. The myriad prospects of biotechnology are even more far-reaching than that of the silicon microchip. The microchip is essentially a device for handling the information, while biotechnology can produce 'materials' from fuels to medicines, from foods to vaccines, and from chemicals to plastics.

Some of the important areas deriving benefit from biotechnology include:

- Human and animal health
- Medicine
- Agriculture
- Horticulture and floriculture
- Forestry
- Dairy
- Fisheries and Aquaculture
- Chemicals and Biochemicals
- Food processing and beverages
- Environment and Pollution control
- Renewable energy and fuels
- Mining
- Crimes and parentage disputes
- Population control by improved contraceptives

Table 1.1   Biotechnology-based companies

Industry	Biotechnological products
1. Therapeutics	Pharmaceutical products for the cure or control of human diseases including antibiotics, vaccines, gene therapy
2. Diagnosis	Clinical testing & diagnosis, food, agriculture
3. Agriculture/ forestry/ horticulture	Novel crops or animal varieties, pesticides
4. Food	Wide range of food products, fertilisers, beverages, and ingredients
5. Environment	West treatment, bioremediation, energy production
6. Chemical intermediates	Reagents including enzymes, DNA, RNA, vitamins, acids, steroids, alkaloids.
7. Equipments	Hardware, software, fermenter, bioreactors, biochips, biosensors

Biotechnology has potential impact on virtually all the aspects of the human welfare. As a result, it plays a very significant role in employment, production, and economics that ultimately improve the quality of human life throughout the world. Indeed biotechnology is destined to be as crucial to the first half of the 21st century as computer science was to the second half of the 20th century.

Biotechnology is a demanding industry that requires a skilled workforce and support from the public so that it can achieve continuous growth. Economics that encourage public understanding and provide a competent labour force should achieve long-term benefits from biotechnology. There can be several categories of companies that can exploit commercial outputs of biotechnology. The major industries involved with biotechnology can be placed in seven categories (table 1.1).

Biotechnology concerns the practical application of organisms or their components. Historically it was an art, involved in the production of wines, beer and cheeses. Nowadays it involves a series of advanced technologies spanning biology, chemistry and process engineering. In recent years innovations involving genetic engineering have had a major impact on biotechnology. Its applications are diverse. Some of the important contributions of biotechnology in relation to human welfare are enlisted in table1.2.

Table 1.2   Some selected applications of biotechnology for the welfare of
             human beings

Area	Applications
Genetic Engineering	Disease diagnosis with the help of monoclonal antibody, DNA probes, immunodiagnostic tools like ELISA, immuno-PCR, etc.
	Prenatal diagnosis of genetical defects.
	Production of antibodies, antibiotics, recombinant vaccines, human insulin (humiline), interferons, human growth hormone, etc. used in the treatment of bacterial and viral diseases and in metabolic disorders.

Table 1.2 *Contd...*

Area	Applications
	Gene therapy for correcting genetic disorders.
	DNA fingerprinting for identification of criminals and solving parentage and other medico-legal disputes.
	Human Genome project for better understanding of gene function.
Industrial biotechnology	Manufacture of fermentation products—ethanol, lactic acid, citric acid, glycerine, acetone, etc. for commercial utility.
	Production of antibiotics—penicillin, streptomycin, erythromycin, cycloheximide, etc. for treatment of microbial disorders.
	Production of immunotoxins for treatment of cancer.
	Production of enzymes like amylase, protease, lipases for use invarious industries and in medicines.
	Production of single cell protein used as human food and animal feed.
	Production of low-cost and efficient biofuel—ethanol, methane, hydrogen, etc. from biomass.
	Immobilised enzymes for industrial applications.
	Extraction of minerals like copper, uranium, etc. from low-grade ores/ solid rock using microbes.
Animal biotechnology	*In vitro* fertilisation (IVF) for developing test tube babies to overcome infertility and other problems of reproductive system.
	Embryo splitting in farm animals for their rapid multiplication with superior genotype.

Table 1.2   *Contd...*

Area	Applications
	Use of gene transfer technique to produce transgenic animals to achieve fast growth, disease resistance and increased milk output and also to act as bioreactors for production of therapeutics.
	Organ culture for regeneration of skin and cartilage utilised in reconstructive surgery.
Plant biotechnology	Micropropagation for ornamental plants, embryo culture for production of haploids, somatic embryogenesis to generate artificial seeds, protoplast fusion for somatic hybrids and cybrids have immense importance.
	Rapid production of commercial fruits and forest trees by clonal multiplication.
	Production of transgenic plants with the help of gene transfer technique to obtain disease, insect and herbicide resistance with improved post-harvest features including seed proteins and fatty acid content.
Environmental biotechnology	Involves the role of genetically engineered microbes in sewage treatment and detoxification of waste and industrial effluents.
	Degradation of petroleum and management of oil spills using bacteria.
	Reducing pollution by avoiding use of chemical pesticides and promoting use of biological entities to control plant disease and insect pests.
	Active contribution of eco-friendly biosensors, bioscrubbers, bioplastics and phytoremediation.

1.4 BIOTECHNOLOGY IN INDIA

The Government of India constituted a National Biotechnology Board (NBTB) to encourage the research activities in biotechnology. In 1986, The Ministry of Science and Technology established a Department of Biotechnology (DBT) that has funded for the establishment of the following centres for Plant Molecular Biology (PMB).

- Madurai Kamraj University, Madurai.

- Jawaharlal Nehru University, New Delhi.

- Tamilnadu Agricultural University, Coimbatore.

- Osmania University, Hyderabad.

- Bose Institute, Calcutta.

- National Botanical Research Institute, Lucknow.

- Delhi University, (South Campus) New Delhi.

In addition to Department of Biotechnology, Council for Scientific and Industrial and Research (CSIR), New Delhi and Indian Council for Agriculture Research (ICAR), New Delhi are also encouraging research activities in Biotechnology. National Bureau of Plant Genetic Resources (NBPGR), New Delhi, Central Institute of Medicinal and Aromatic plants (CIMAP), Lucknow and Tropical Botanical Garden and Research Institute (TBGRI), Trivandrum, are the three institutes that have gene bank facilities for medicinal and aromatic plants.

Micropropogation of the forest and fruit plants is carried out in the following three institutes with financial support from DBT.

1. Tata Energy Research Institute (TERI), New Delhi.

2. National Chemical Laboratory, Pune.

3. J.N. Vyas University, Jodhpur.

DBT has also started an extensive programme for development of manpower in biotechnology. This programme consists of master level training at 19 centres located in the different universities all over India including Department of Biotechnology, Pune University, Pune and School of life sciences, North Maharashtra University, Jalgaon.

In addition, some private companies have developed interest in biotechnological research in North Maharashtra region because of its suitable climates, soil and other environmental conditions. A single private sector belonging to JB group has pioneered in tissue culture technique for banana plant and is flourishing its business for the past four decades not only in India but also in foreign countries. Some other companies are manufacturing diagnostic kits for certain diseases and essential components for DNA fingerprinting.

Review Questions

1. Define biotechnology in various forms.

2. Comment on the interdisciplinary nature of biotechnology.

3. Discuss the scope and importance of biotechnology in promoting human welfare.

4. Enlist the benefits derived from biotechnology.

5. What are the seven categories of companies based on biotechnology?

6. Write short notes on:

    a. Department of Biotechnology (DBT)

    b. Medical biotechnology

    c. Industrial biotechnology

    d. Animal and plant biotechnology

    e. Environmental biotechnology

2

ANIMAL CELL AND TISSUE CULTURE

2.1 INTRODUCTION

The cell is a unit of biological activity delimited by a semi-permeable membrane and capable of self-reproduction in a suitable non-living medium. The organisms made up of only one cell are called unicellular organisms while those having many cells in their body are called multicellular organisms.

The term **tissue culture** means the culture of whole organs, tissue fragments as well as dispersed cells on a suitable nutrient medium. Tissue culture can be divided into a) **Organ culture** and b) **Cell culture**. In **organ culture** whole embryonic organs or small tissue fragments are cultured *in vitro* in such a manner that they retain their tissue architecture. **Cell cultures** on the other hand are obtained either by enzymatic or mechanical dispersal of tissue into individual cells or by spontaneous migration of cells from an explant, and they are maintained as attached monolayers or as cell suspension.

Culture media are nutrient media used for culture of animal cells and tissues. It must be able to support their survival as well as growth. Freshly isolated cell cultures are called **primary cultures**. Once a primary culture is subcultured, it gives rise to **cell lines**. When a complete animal is obtained from somatic cells of an animal, it is called **animal cloning**. In 1997, **Dolly the sheep** was created by Roslin Institute, UK and it was the first mammal to be cloned from an adult cell.

Cells derived from a single cell through mitosis constitute a **clone** and the process of obtaining clones is called **cloning.** In other words asexual progeny of a single individual make up a clone.

Animal cell cultures are used to generate valuable products like virus vaccines and biochemicals which are mainly high molecular weight proteins like enzymes and hormones, cellular biochemicals like interferon and immunological compounds like monoclonal antibodies. Transplantable tissues and organs are other valuable products derived from cell culture. Artificial skin and transplantable cartilage have great significance in curing many disorders.

2.2 CULTURE MEDIA

The nutrient media used for culture of animal cells and tissues are called culture media. They must provide nutritional, hormonal and stromal factors for the survival and growth of the cultured cells and tissues.

2.2.1 Types of Culture Media

There are two categories of culture media: a) Natural media and b) Artificial media. The selection of medium depends on the type of cells to be cultured (normal, immortalised or transformed) and the objective of culture (growth, survival, differentiation or production of desired proteins).

Natural media These media include naturally occurring biological fluids derived from organisms and are of four types (i) blood plasma (ii) blood serum (iii) tissue extract and (iv) complex natural media.

i. **Blood plasma** Plasma provides the complete nutrients in which cells can survive, grow and multiply. It is obtained by centrifugation of whole blood before coagulation can take place. Plasma is now commercially available either in liquid or lyophilised state. After placing the tissue in plasma, coagulation is encouraged by the addition of a small amount of tissue extract or thrombin because the cells in culture require a solid support for continued growth and activity.

ii. **Blood serum** It is blood plasma without fibrinogen. It may be obtained from adult human blood, placental cord blood, horse blood or calf blood. The serum provides some of the growth factors or some of the physical conditions, or both; they are not found in artificial media. Serum preparations must be tested for sterility and toxicity before use.

iii. **Tissue extracts** Chick embryo tissue extract has remarkable powers of promoting cell growth and multiplication in cultures of connective tissue cells.

iv. **Complex natural media** Some of them are as follows:

Supplemented Hanks–Simms medium

Composition: Hanks–Simms solution, beef embryo extract, horse serum, and antibiotics.

Supplemented bovine amniotic fluid medium

Composition: Bovine amniotic fluid, horse serum, bovine embryo extract, Hanks balanced salt solution and antibiotics.

Serum-supplemented yeast extract medium

Composition: Yeast extract in the form of Filco's yeastolate, glucose solution, Hanks balanced salt solutions, human serum and sodium bicarbonate.

Serum-supplemented lactalbumin hydrolysate and yeast extract medium

Composition: Earle's saline containing lactalbumin hydrolysate, yeast extract and human or ox serum.

Artificial media   Contain inorganic and organic constituents and may also contain amino acids, fatty acids, vitamins, etc. Some of the synthetic media are as follows:

Medium No.612   Contains cysteine, glutathione and ascorbic acid.

Medium No. 635   Contains purines, pyrimidines as well as adenosine triphosphate, adenylic acid, ribose and deoxyribose.

Medium No. 858   Includes deoxyribonucleosides, coenzymes, and sodium glucuronate in addition to medium No. 635. This medium contains only L-form amino acids.

Medium No.866   It is similar to medium No. 858 supplemented with three fatty acids namely linoleic acid, linolenic acid and arachidonic acid.

Medium CMRL-1066   It is identical to the medium No. 858 supplemented with vitamin B-complex and parahydroxy benzoate.

The above-mentioned synthetic media are free from serum hence they are also known as serum-free media in which the toxic effects of serum are avoided and there is no danger of degradation of sensitive proteins by serum proteases.

2.3 PRIMARY CULTURE AND CELL LINES

It consists of cells freshly removed from an animal and placed in a dish with cell culture medium. Most primary cultures of animal cells are obtained from embryos, whose tissues are more readily dissociated into single cells than those of adults. To accomplish the dissociation, the tissue is removed from the embryo and usually treated with a proteolytic enzyme such as trypsin, which is capable of digesting the proteinaceous material that helps hold the cells together. The tissue is then washed free of the enzyme and usually suspended in a saline solution that lacks Ca^{2+} ions and contains a substance such as ethylene diamine tetraacetate (EDTA), which binds (chelates) calcium ions.

Once the cells have been dissociated into a cell suspension, they can be cultured directly or can be separated according to cell type and then placed in culture. Once the cells have been separated, there are basically two types of primary cultures that can be started. In a mass culture, a large number of cells are added to a culture dish, they settle and attach to the bottom and form a relatively uniform layer of cells. These cells that survive will grow and divide and, after a number of generations, form a monolayer of cells that covers the bottom of the dish (figure 2.1).

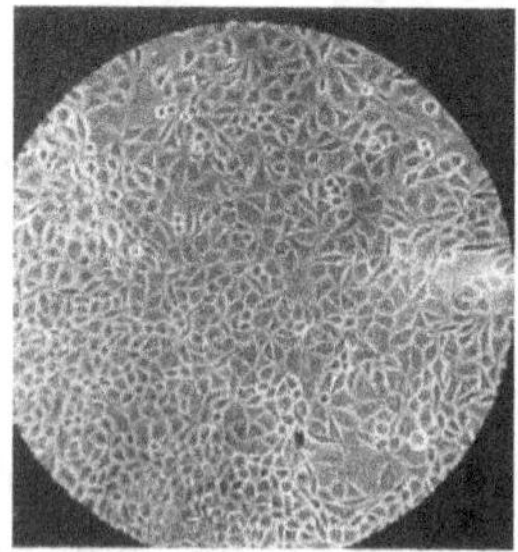

Figure 2.1 Mass cultures of cells forming a monolayer on the surface of culture dish

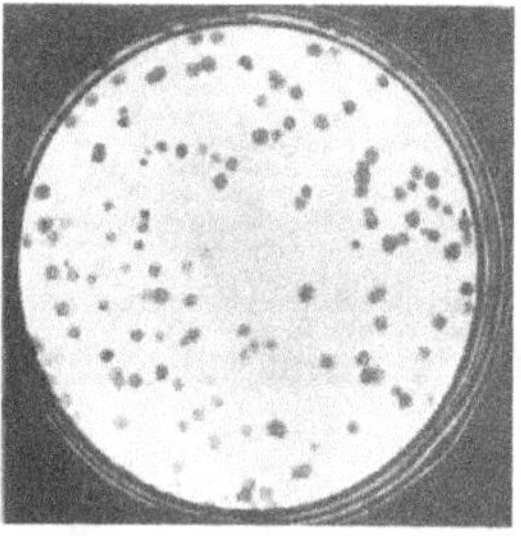

Figure 2.2 Clonal culture of the cells showing scattered colonies over the surface of culture dish

In the other type of culture, called a **clonal culture**, a relatively small number of cells are added to the dish, each of which after settling and attaching to the surface, is at some distance from its neighbours. In this case, the proliferation of the cells generates individual colonies or **clones** of cells (figure 2.2) whose members are all derived from the same original cell. Thus cells derived from a single cell through mitosis constitute a **clone** and the process of obtaining clones is called **cloning.**

Normal (nonmalignant) cells are capable of a limited number of cell divisions (typically 50 to 100) before they undergo a process of senescence and death. Because of this, many of the cells that are commonly used in tissue culture studies have undergone genetic modification that allow them to grow indefinitely. Cells of this type are called a **cell line**, and they are capable of growing into malignant tumours when injected into susceptible laboratory animals. Thus cell lines are raised by sub culturing the primary culture. When cell lines die after several subcultures they are called **finite cell lines** or they may continue to grow indefinitely to form **continuous cell line**. Usually normal tissue gives rise to finite cell lines, while tumours give rise to continuous cell lines. But there are several examples of continous cell lines which were derived from normal tissues and are themselves nonmalignant e.g. MDCK dog kidney, mouse 3T3 fibroblasts, etc. The evolution of continous cell lines from primary cultures is supposed to involve a mutation that alters their properties as compared to those of finite lines.

2.4 DOLLY THE SHEEP—THE FIRST CLONED MAMMAL

The announcement of the birth of Dolly at Roslin institute, UK in 1997 attracted huge media attention, because she was the first mammal to be cloned from an adult cell. Before this, researchers had cloned a tadpole from an adult frog, and sheep and cattle from early embryo cells, but most scientists believed it could not be done from an adult cell because adult cell had specialised roles that could not be altered. Ian Wilmut, Keith Campbell and colleagues of the Edinburgh Roslin Institute using a technique called **nuclear transfer,** successfully created Dolly (figure 2.3).

The method is illustrated in the figure 2.4 and involves transferring the nucleus of a normal cell (containing all the donor's genetic material) to an unfertilised egg

Figure 2.3 Dolly, the first cloned mammal

from which the maternal nucleus (with all its genetic material) has been removed. The fused nucleus and egg are then 'activated' with a short electric impulse. Because the resulting 'reconstructed embryo' has only one parent (the cell's donor), it is genetically identical to that parent, or it is a clone. Despite the triumph of Dolly, success rates are still low in all species, with less than 1% of such transfers leading to live birth.

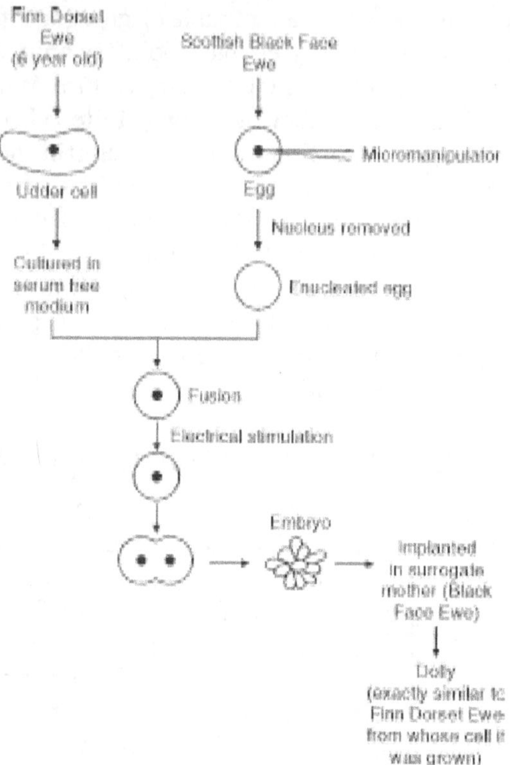

Figure 2.4 Steps involved in the development of Dolly

2.4.1 Applications of Animal Cloning

There will be many future benefits from the technology that produced Dolly (It could in principle, be used to create an infinite number of clones of the very best farm animals). Once it has been shown that such clones are healthy and that the technology can be applied without compromising on animal welfare. And because it is a means of converting cells in culture to live animals, it also provides an ability to introduce precise genetic modifications into farm animal species.

This would have valuable potential for the efficient production of human therapeutic proteins which are in great demand for the treatment of many diseases. The output of such protein in the milk of transgenic sheep, goats and cattle can be as high as 40 g per litre.

Alpha-1-antitrypsin is one such protein produced by this technology, it is currently undergoing advanced clinical trials for the treatment of cystic fibrosis and emphysema.

There are also implications for the cell transplants that are being developed for a variety of common diseases including Parkinson's disease, heart attack, stroke and diabetes. The cloning of adult animals from a variety of cell types shows that the egg and early embryo are able to 'reprogramme' an adult cell. Understanding more about the mechanism involved may lead to a way of reprogramming a patient's own cells so they will not be rejected, without using embryonic stem cells.

2.5 ORGAN CULTURE

It is the process of *in vitro* culture and growth of organs or parts thereof in such a way that their structure and function resemble closely the concerned organs *in vivo*. Although this technique has been used to grow intact embryonic organs, the name is actually something of a misnomer since it is usually employed to maintain small pieces of organs *in vitro*. The object of the organ culture technique is to maintain the architecture of tissue and to direct it towards normal development so that it occurs *in vitro*. In order to achieve this aim, it is essential that the tissue should never be disrupted or damaged and this requires careful handling, so that organ culture techniques generally demand more careful manipulation than tissue culture technique.

2.5.1 Techniques in Organ Culture

The techniques of organ culture can be divided into those employing a solid medium and those employing a fluid medium. It may utilise one of the following types.

i. Plasma clot   It is the *classical technique of organ culture*, in which the explant is cultured on the surface of a clot consisting of chick plasma and chick embryo extract contained in a watch glass. This technique is used for studying morphogenesis in embryonic organ rudiments; it has been also modified to study the action of hormones, vitamins, carcinogens, etc. on adult mammalian tissues (figure 2.5).

Figure 2.5 A watchglass method for culture explant on a plasma clot

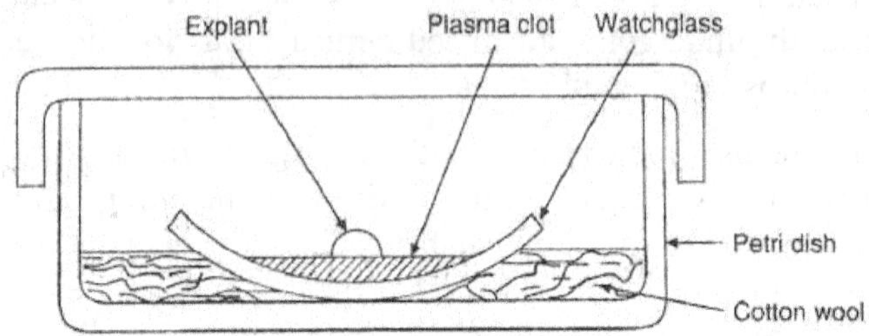

Figure 2.5   A watchglass method for culture explant on a plasma clot

ii. Raft Method   In this method the explant is placed onto a raft of lens paper or rayon acetate that is floated on serum in a watch glass. Treating it with silicon enhances the floatability of lens paper. Usually four or more explants are placed on each raft.

iii. Agar gel Method   In this method, the medium is gelled with 1% agar. It avoids immersion of explants into the medium and permits the use of defined media. The agar gels are generally kept in embryological watch glasses (figure 2.6) and sealed with paraffin wax. This method is used to study many developmental aspects of normal organs as well as of tumours.

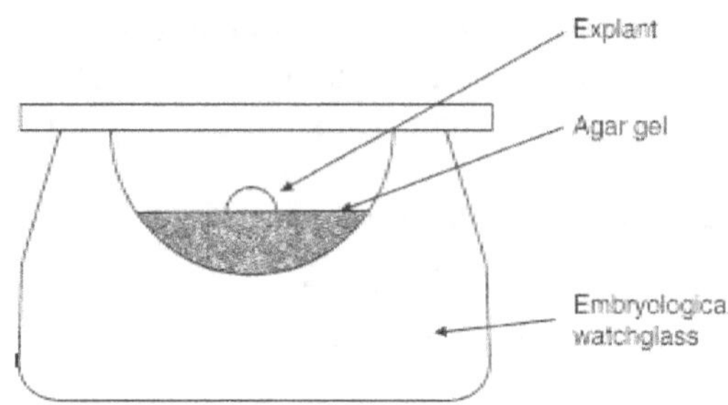

Figure 2.6 Organ culture in agar gel

iv. Grid Method   In this method a suitable wire mesh or perforated stainless steel sheet whose edges are bent to from four legs of about 4mm height is used. The tissues derived from cartilage and bones are kept directly on the grid but softer tissues like glands or skin are first placed on raft (made of lens paper or rayon acetate) which are then kept on the grids (figure 2.7).

Using this approach the growth and differentiation of embryonic and adult tissues can be studied.

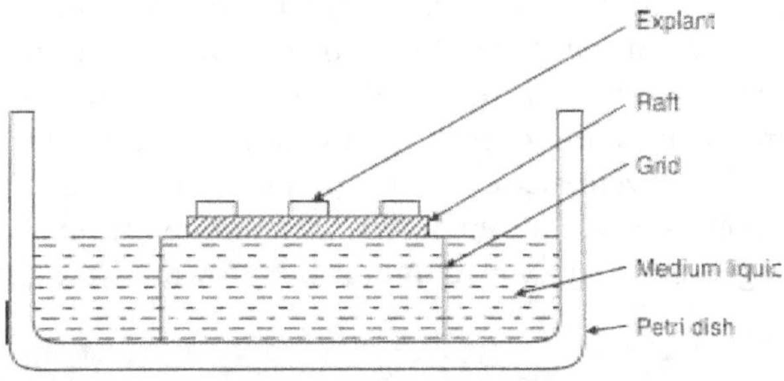

Figure 2.7   Organ culture by grid method

v. Cyclic exposure to medium and gas phase This technique is successfully used for long-term culture of human adult tissues. The explants are exposed rhythmically to the fluid medium and gas phase. The explants are attached to the bottom of a plastic culture dish and are covered with fluid medium. The dishes are enclosed in a chamber containing a suitable gas mixture and mounted on a rocker platform. The chamber is rocked at several cycles/min to ensure intermittent exposure of the organ explants to medium and gas phases.

2.5.2 Applications

Organ culture is essentially a technique for studying the behaviour of integrated tissues rather than isolated cells. They have been widely used for the following:

1. They are used to study the pattern of growth, differentiation and development of organ rudiments and the effects of various factors like antibiotics, hormones, vitamins, etc. on the above-mentioned parameters.

2. *In vitro* study of the action of drugs and carcinogenic agents on the animal organ serves as an index for the whole animal.

3. The most fabulous application of organ cultures is to produce tissues for implantation in patients; this is often known as tissue engineering. Human skin and cartilage have been successfully produced *in vitro* and used for transplantation.

2.6 ARTIFICIAL SKIN

It is now possible to grow skin by *in vitro* technique, in fact only the epidermis portion of skin is regenerated in culture. Keratinocytes derived from any type of stratified squamous epithelium, those of the epidermis, cornea, mouth, tongue, etc. will divide *in vitro* to form colonies and eventually a continuous epithelium (figure 2.8). With improvements in the techniques it is possible to reconstitute complete skin—both epidermis and dermis. It involves addition of collagen matrix in a medium for the growth of the tissue.

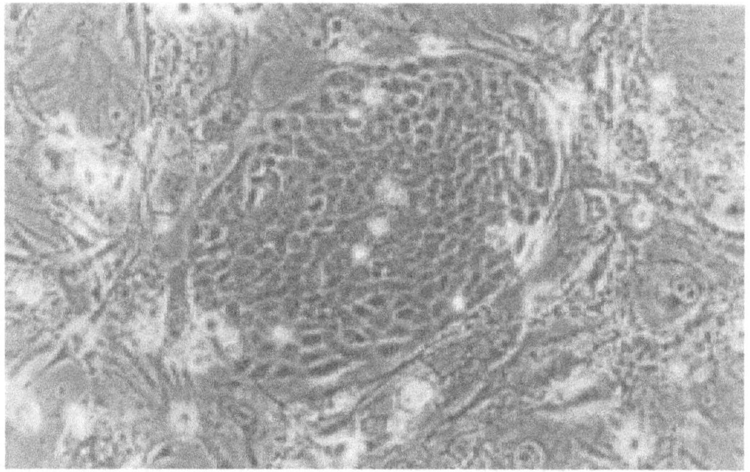

Figure 2.8 *In vitro* culture of keratinocytes (human epidermal cells) to form a continuous epithelium

For regenerating artificial skin, the skin explant may be obtained from the patient's body or from prepuce (the loose skin at the tip of penis) of newborn babies. Skin cells of newborns grow faster than those of adults; the use of a synthetic polymer called PGA allows the newborn skin to grow without scars. This skin can be used to cover the wound till the patient's skin is cultured and artificial skin obtained for grafting.

Treating the skin explant with trypsin dissolves the bulk of epidermis called keratinocytes. These cells are cultured in the vessels, the bottom of which is covered with irradiated 3T3 fibroblast cell line for increasing the proliferation of keratinocytes. They grow to form colonies, which are again dissociated into single cells and cultured in the same manner. The process is repeated till a sheet of pure epithelium is formed. This sheet is detached from tissue culture vessels, enzymatically or by reducing level of vitamin A in the medium; it is cleaned and used for grafting.

The ability to lift up sheets of cultured skin cells and graft then led to the remarkable use of a tissue culture grown skin to treat people with severe burns and chronic skin ulcers. Because skin cells grow so well in culture, it is possible to take a small piece of an individual's skin and within 2 to 3 weeks, grow enough skin in the laboratory to cover the entire body. A 3 cm^2 skin explant can yield about 1.7 m^2 artificial skin in 3–4 weeks. Since the skin is derived from the patient's own cells, there is no danger of rejection of graft. However, the regenerated skin is not completely normal, since it lacks hair follicles and sweat glands.

2.7 CARTILAGE REGENERATION

Artificial cartilage developed *in vitro* by culturing the human chondrocytes has promising role in human implantation particularly in the cases of injuries, arthritis, etc. In 1966, Coon and Cahn described a technique for cultivation of cartilage- synthesising cells from chick embryo somites. Cahn and Lasher (1967) later used this system for analysis of the involvement of DNA synthesis as a prerequisite for cartilage differentiation.

Cartilage forming cells, chondrocytes, are embedded in a dense matrix of proteoglycans, which must be digested by sequential enzymatic treatment to release small cellular compartments of the tissue. The cells are then cultured in slightly alkaline Ham's medium with serum and increased Mg^{2+}.

Further improvement in this technique involves use of synthetic biocompatible and biodegradable polymers seeded with chondrocytes and implanted to regenerate cartilage. It seems that synthetic biocompatible and biodegradable materials could be a useful alternative as a template for cartilage regeneration. These materials can be altered in many different ways if required and are eventually completely replaced by cartilage.

Cartilage replacement or regeneration is utilised mainly in reconstructive surgery, such as the nose and ears, and in orthopaedic surgery for the correction of joint problems such as those caused by arthritis.

Key Concepts

□ Tissue culture is the method of growing an entire organ or parts on a suitable nutrient medium in aseptic conditions.

□ Organ culture involves *in vitro* technique for whole embryonic organs or small tissue fragments, whereas cell cultures are used to generate valuable products like vaccines, enzymes, hormones, interferons and antibodies.

□ Nutrient media used for culture of animal cells and tissues may be natural or synthetic.

□ Primary cultures are freshly isolated cell cultures which when subcultured, give rise to cell lines.

□ The cells derived from a single cell through mitosis constitute a clone.

□ Dolly, the sheep, was the first mammal cloned from an adult cell. Cloning provides an ability to introduce precise genetic modifications into farm animal species for production of human proteins that are in great demand for treatment of many diseases including Parkinson's disease, heart attack, stroke and diabetes.

□ Artificial skin and cartilage replacement or regeneration has a promising role in human implantation particularly in the cases of injuries caused by burns, skin ulcers and in constructive and orthopaedic surgery.

2.8 ANIMAL CELL FUSION

Hybrid cells can be formed by the fusion of two different types of somatic cells. The hybrid cells thus produced have several important uses which may be listed as follows:

1. in gene or chromosome mapping

2. to study the control of cell division and pattern of expression of gene

3. in production of monoclonal antibodies by producing hybridoma

4. in providing ideas about malignant transformation

5. to obtain viral replication

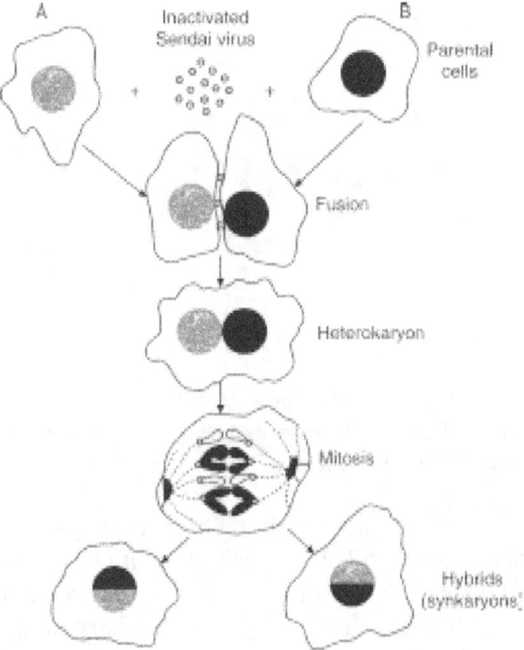

Figure 2.9   Sendai virus mediated cell fusion to produce heterokaryon
and its hybrid cell line

Cells can be fused through the use of inactivated Sendai virus (a member of the parainfluenza virus) and other agents that affect membrane structure, such as polyethyleneglycol (PEG) and lysolecithin. These agents adhere to the plasma membrane of cells and alter their properties in such a way that facilitates their fusion. Fusion of two cells produces a single hybrid cell with two nuclei. This is the initial product of fusion called **heterokaryon** as shown in figure 2.9. Subsequently, both nuclei might enter mitosis synchronously, form a single metaphase plate, divide, and produce a hybrid cell line (also called a synkaryon). The cells of a hybrid cell line have single nucleus containing chromosomes from both parental nuclei.

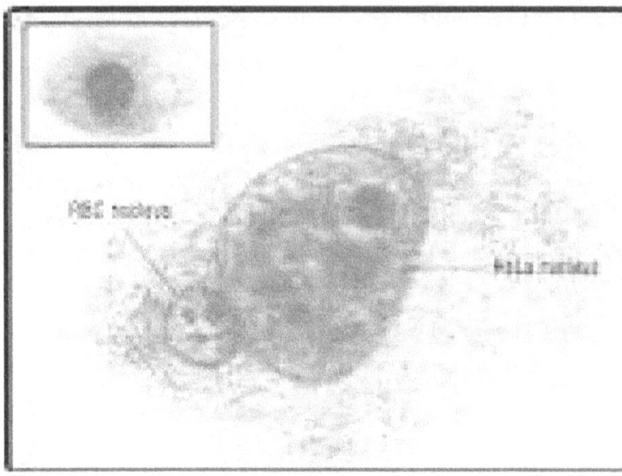

Figure 2.10   Heterokaryon formed by fusion of chick red blood corpuscle with a
human HeLa cell. The inset shows a normal chick erythrocyte with
condensed inactive chromatin.

In 1965, H. Harris found that chick erythrocyte nuclei are
reactivated when fused to HeLa cells (an undifferentiated cell line so
called because it was derived from the uterine carcinoma cells of a
woman named Henrietta Lacks). These heterokaryons are of interest
because the erythrocyte nucleus does not normally synthesise RNA
or DNA. Chick erythrocytes are terminally differentiated cells that
have a highly condensed nucleus and are destined to die
(figure 2.10). (Mammalian erythrocytes normally eliminate their
nuclei during red blood cell maturation, whereas red blood cells of
birds amphibians, and reptiles retain their nuclei which became
inactivated). When fused to HeLa cells, the chick erythrocyte nucleus
increases 20 times in volume, disperses its chromatin, resumes RNA
synthesis, develops a nucleolus, and eventually replicates its DNA.
This process is accompanied by the uptake of large amounts of human
nuclear proteins, which are thought to reactivate the erythrocyte
nucleus.

These experiments clearly show that the cytoplasmic environment
controls the synthesis of macromolecules in a nucleus. Even though
the erythrocytes are terminally differentiated cells, they can resume
RNA and DNA synthesis.

In a few cases it has been possible to activate latent genes coding for specific proteins by fusion of differentiated cells for example, albumin secreting rat hepatoma cells fused with mouse lymphocytes (which do not produce albumin) will occasionally give rise to clones of hybrid cells that are able to produce both rat and mouse albumin. Somatic cell hybridisation experiments show that substances present in cell cytoplasm can modify the patterns of nucleic acid synthesis and gene expression by a nucleus. In one such experiment, the frog oocyte cytoplasm is able to reprogramme the expression of genes in transplanted nuclei. This reprogramming ability was found when kidney nuclei of *Xenopus* were injected into oocytes of the salamander *Pleurodeles*. The work with somatic cell fusion and transplantation of nuclei into oocyte suggests that the cytoplasm can indeed reprogramme the activity of nuclear genes.

2.9 SOMATIC CELL FUSION AND CHROMOSOME MAPPING

Chromosome mapping is based on the somatic cell hybridisation in which the whole genome of a cell is transferred into another by a parasexual mechanism. Experimentally a cultured human cell is fused with a mouse cell, by the action of fusogenic agents such as Sendai virus or polyethylene glycol. In subsequent cell divisions these hybrid cells tend to lose human chromosomes (figure 2.11). The experiment started with parent cells having 3 chromosomes. After the fusion and subsequent loss of human chromosomes, clones of the cells containing chromosome 1, 2 or 3 are produced. The presence of specific gene products for each chromosome is demonstrated by a sensitive method that involves the separation of isoenzymes by electrophoresis. The isoenzymes of humans and mouse may have different electrophoretic mobility due to differences in the charges of amino acids in the polypeptides.

Another useful experiment consists in selecting hybrid cells that retain a specific human chromosome. This may be done using mutated mouse cells which are unable to grow in a particular medium unless their genetic enzyme deficiency is compensated by the human gene.

A cell uses two biochemical pathways to produce the nucleotides needed for DNA replication. The *de novo* pathway, in which sugar

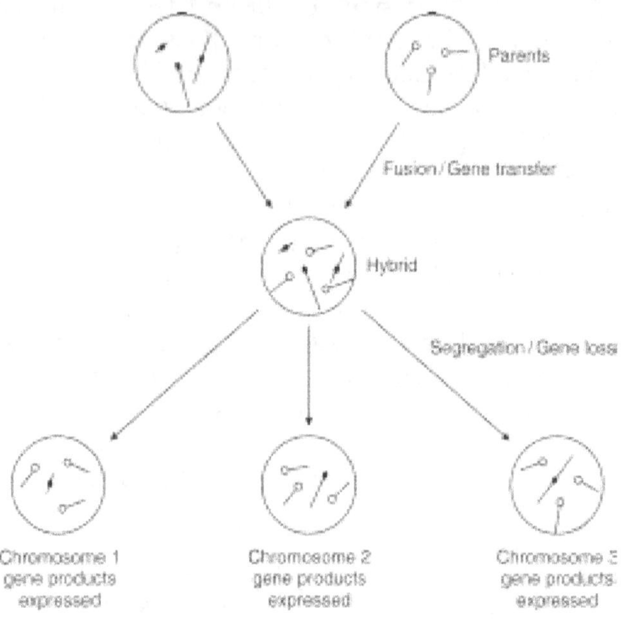

Figure 2.11   Diagram showing the role of somatic cell fusion in chromosome
               mapping

and amino acids are used, can be blocked with the aminopterin
contained in the HAT medium (which contains hypoxanthine,
aminopterin and thymidine). Hypoxanthine (a precursor of purine)
and thymidine (a precursor of pyrimidine) are used for the synthesis
of nucleotides by salvage pathway, provided that the enzymes
thymidine kinase (TK) and hypoxanthine guanine phosphoribosyl
transferase (HGPRT) are present. When TK-deficient or HGPRT-
deficient mouse cells are hybridised with human cells, the genes
contained in human chromosomes can complement the deficiency,
compensating for the defect. Eventually clones of cells produced on
HAT medium have one chromosome containing TK gene (present
in human chromosome No. 17). Once this is achieved, it is possible
by gel electrophoresis to identify isoenzymes and other biological
markers such as structural proteins or surface antigens, encoded by
this human chromosome.

Thus genes incorporated by somatic cell hybridisation are transcriptionally and translationally active. In other words they are able to produce mRNA and the corresponding specific protein, which can be used as markers for the localisation of gene in the corresponding chromosome and hence forming one of the useful methods for gene mapping.

2.10 SOMATIC CELL HYBRIDS PRODUCING MONOCLONAL ANTIBODIES

The role of cells in the immune response is witnessed by production of antibodies in the serum when animals are injected by foreign substances in the form of a protein or complex polysaccharide (known as antigen) that usually reside on surface of invading microorganisms. Antibodies are protein molecules that bind specifically to the foreign antigen. Many different antibodies appear in the serum of an immunised animal, each one recognising a different part of the antigen's shape. Furthermore, individuals of the same species have different immunological responses, so that two antisera directed against the same antigen can in fact be very different. This is a major problem in medicine because, for example, success or failure of organ transplantation depends on whether the donor and the recipient patient have the correct histocompatibility antigens on the cell surface. Antibodies can be used as standardised diagnostic reagents. It is possible to create cell lines that produce only a single type of antibody by cell fusion technique. Lymphocytes produce antibodies as a counteracting response. Each antibody-producing cell can synthesise only one type of antibody. Lymphocytes do not multiply in culture, but in 1976, G.Kohler and C. Milstein in England developed a technique by which a single antibody-producing cell can be multiplied indefinitely in culture by hybridising with a tumour cell.

The process of monoclonal antibody production starts with a mouse that is immunised against the desired antigen. The lymphocytes from the spleen of this mouse are then fused with a mouse cell line derived from the tumours found in myelomas (cells invading bone marrow). The specificity of myeloma cell is that it carries a mutation in the enzyme hypoxanthine-guanine phosphoribosyl transferase (HGPRT) and is therefore unable to grow in HAT medium (culture medium containing hypoxanthine, aminopterin and thymidine).

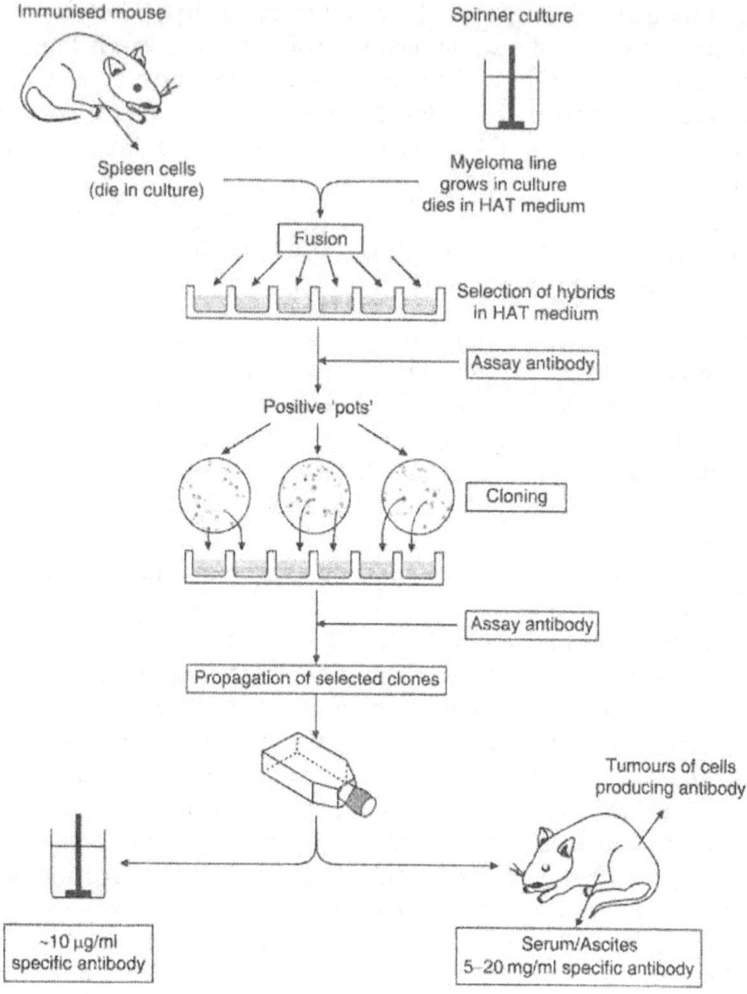

Figure 2.12 Steps in the production of monoclonal antibody

However the resulting hybrid myeloma spleen cells (hybridoma) express both the lymphocyte's property of antibody production and the myeloma's characteristic of continued cell division. Individual hybridoma cells can be isolated, cloned in small wells, and tested for antibody production (figure 2.12).

Key Concepts

- Experiments in which a nucleus is placed in a foreign cytoplasm have provided insights on how genes are regulated.

- Cell fusion by Sendai virus allows a nucleus to be placed in a different cytoplasm. A single cell with two nuclei is called a heterokaryon.

- Cell nuclei that no longer synthesise RNA or DNA (e.g. chicken erythrocytes) increase in volume, resume RNA synthesis and may replicate their DNA when fused with HeLa cells. Reactivation of nucleus is probably related to the entrance of proteins.

- Fusion of B-lymphocytes with myeloma cell lines permits the production of immortal cell lines that secrete highly specific monoclonal antibodies.

- The work with somatic cell fusion and with transplantation of nuclei into oocytes suggests that the cytoplasm can indeed reprogramme the activity of nuclear gene.

- Human genes can be transferred to a mouse cell by somatic cell hybridisation.

- A selection (HAT) system may be used to seggregate mutations and clones of cells—one of the best methods for gene mapping.

In addition to this, individual clones of antibody producing hybridomas can be reinjected into animals, where they reproduce themselves to an ever-greater degree. Cells producing high titers of a desired antibody can be frozen in liquid nitrogen, defrosted and regrown whenever particular antibodies are needed. Because all of the cells of one clone are derived from a single lymphocyte, a monoclonal antibody of high purity is produced. Monoclonal antibodies have immense significance in human health (details are given in section 5.3). The cell fusion technique eventually produces immunological tools of practical importance to all mankind.

2.11 MICROPROPAGATION

Micropropagation is the tissue culture technique used for rapid vegetative multiplication of ornamental plants and fruit trees by using small-sized explants. Propagation of heterozygous genotypes by vegetative multiplication is of great necessity as it involves only mitotic cell division. Progeny obtained by vegetative propagation or asexual reproduction of a single plant or individual constitutes a **clone.** The members of a single clone have the same genotype. This micropropagation is also known as **clonal propagation**. It is the only process adopted by Indian plant biotechnologists in different industries mainly for the commercial production of ornamental plants like lily, orchids, chrysanthemum, eucalyptus, feijoa, cinchona, blueberry, etc. and fruit trees like banana, grapes, potato, citrus, oil palm, etc.

The method of micropropagation involves the following steps:

1. Selection and sterilisation of suitable plant parts or organs, mainly axillary buds or adventitious buds, bulbs or protocorms, etc. and their inoculation into suitable nutrient medium. The selected part for culture is the initial explant or micropropagule.

2. The initial explant is multiplied on multiplication medium (it contains naphthalene acetic acid (NAA) and higher content of salt).

3. This medium induces multiplication of explant, i.e. large numbers of similar shoot tips are formed.

4. Then these shoot tips are separated and transferred to regeneration medium to form plantlets. (This medium contains growth hormones but is without NAA and has less content of salt).

5. Then the plantlets are transferred to soil to form complete functional plants.

2.11.1 Advantages of Micropropagation

1. It helps in rapid multiplication of plants.

2. A large number of plantlets are obtained within a short period and from a small space.

3. Plants are obtained throughout the year under controlled conditions, independent of seasons.

4. Genetically similar plants (clones) are formed by this method. Therefore desirable characters (genotype) and desired sex of superior variety are kept constant for many generations.

5. The rare plant and endangered species are multiplied by this method and such plants are saved.

6. Sterile plants or plants which cannot maintain their characters by sexual reproduction are multiplied by this method.

7. It is an easy, safe and economical method for plant propagation.

8. In case of ornamentals, tissue culture plants give better growth, more flowers and less fall-out.

2.12 SOMATIC EMBRYOGENESIS (S.E.)

In plant tissue culture medium a single somatic plant cell forms an undifferentiated mass of cells called **callus**. Small, bipolar, embryo-like structures called somatic or non- zygotic embryoids develop at the callus surface and have a number of structural similarities to the sexual or zygotic embryos found in the developing seeds. They can, under appropriate conditions, be grown on a standard maintenance medium into whole plants (figure 2.13). Thus somatic embryogenesis is an alternative pathway of whole plant regeneration observed in cell culture of some plant species. But it is rarely used and is successful in the case of carrots, celery and alfalfa. Somatic embryogenesis may be direct (without formation of callus) or indirect (involving formation of callus from the explant). Somatic embryogenesis is greatly affected by the presence of chemical factors like auxin, cytokinine, gibberllin, etc. in the growth medium. In addition, substantial amount of reduced nitrogen, optimum level of dissolved oxygen and high potassium in the medium supplemented with activated charcoal have facilitated embryogenesis in several cultures.

The mass production of adventitious embryos in cell culture is an ideal propagation system. Somatic embryos have no food reserve but suitable nutrients could be packed by coating or encapsulation to form some kind of artificial seeds.

The artificial seeds are significant in the following ways.

1. Production of artificial seed is independent of the reproductive phase of plant.

2. The plant can be produced at any season of a year.

3. The plants with shortened life cycle can be obtained by growing artificial seeds.

4. Artificial seeds give protection to meiotically unstable, elite genotype.

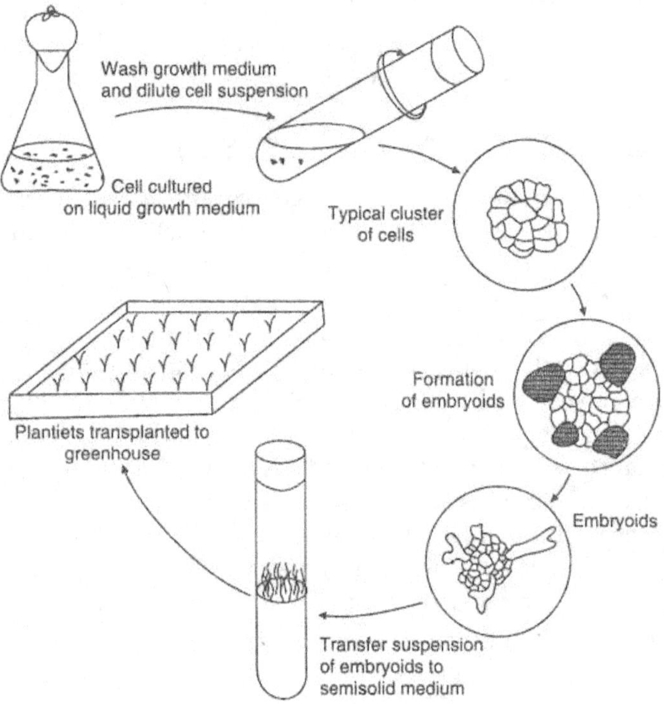

Figure 2.13 Illustration of somatic embryogenesis in culture

2.13 PROTOPLAST FUSION TECHNIQUE

The successful establishment of *in vitro* culture of plant tissues or organs can be possible only when its cell walls are stripped off, because cell wall is the only barrier between the environment and the interior of the cell and poses a problem for varieties of experimental manipulations. Therefore the plant scientists made an intelligent attempt to remove the wall of plant cell and culture the naked cell on the nutrient medium. After the refinement of methodology plant protoplast culture became successfully established as a novel technique in the field of plant tissue culture.

Every plant cell has a definite cellulosic cell wall and the protoplast lies within the cell wall. The protoplast is a part of a plant cell which lies within the cell wall and can be plasmolysed and isolated by removing the cell wall by mechanical or enzymatic procedure. Therefore, isolated protoplast is only naked plant cell surrounded by plasma membrane. When it is placed in culture medium it has the capacity of cell wall regeneration, cell division and plant regeneration. The plant protoplast either from somatic cell of the same plant or from two genetically different plants can fuse *in vitro* to form somatic hybrid or cybrid which have a great significance in plant genetics.

2.13.1 Sources of Protoplast

Protoplasts can be isolated from varieties of plant tissues. The most convenient and suitable materials are leaf mesophyll, root, stem cortex, hypocotyls, endosperm, petal, coleoptile, pollen mother cell and cells from liquid suspension culture.

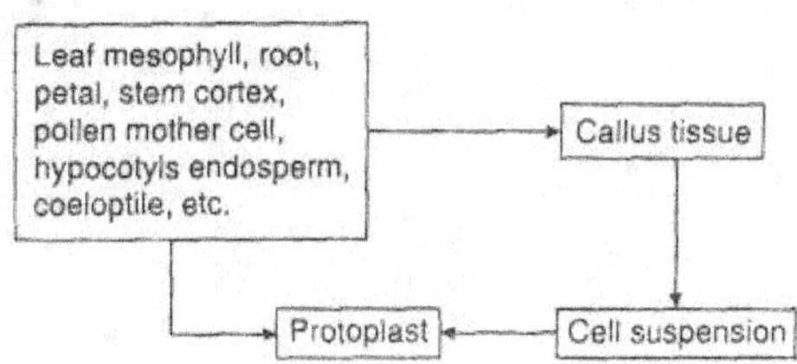

2.13.2 Isolation of Protoplast

During isolation of protoplast, the osmotic pressure should be considered as the most influencing factor since the removal of cell wall leads to an increase in the outward pressure and the expression of protoplast causes it to burst. Therefore, if the cell wall is to be removed to isolate the protoplast, the cell or tissue must be placed in a hypertonic solution of mannitol or sorbitol to provide a stable osmotic environment for the protoplast. Once the cells are stabilised, the protoplast can be isolated from the cell either mechanically or enzymatically.

The operation involved in mechanical isolation of protoplast can be done carefully on small pieces of plant tissue under a microscope using a microscalpel. A lot of time and effort are required to liberate even a few protoplasts. While the most efficient way of isolating protoplasts is to digest the cell wall using cell wall degrading enzymes such as cellulase, pectinase, hemicellulase or micerozyme, some commercially available enzyme preparations like Driselase, Pectolyase Y-23, Onozuka R-10, and Rhizome HP150 can also be used. Factors affecting the process of isolation of protoplasts like pH, temperature, osmotic concentration and incubation time should be adjusted accordingly to stabilise the protoplast and prevent its bursting (figure 2.14).

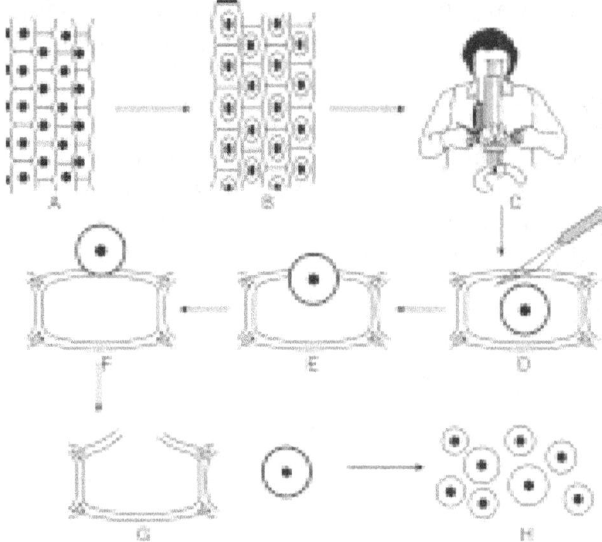

Figure 2.14 Steps involved in mechanical isolation of protoplast (A to H)

Viability of protoplasts after isolation and during culture in liquid medium is very important. The viability of protoplast can be tested by 0.01% fluorescein diacetate or phenosafranin.

2.14 PROTOPLAST FUSION AND SOMATIC HYBRIDISATION

The significant achievement of protoplast culture lies in the possibilities of fusion of one protoplast with another. Normally, isolated protoplasts do not fuse with each other, but their fusion can be induced by chemicals like polyethylene glycol (PEG) to produce a hybrid protoplast which further regenerate a cell wall to produce a somatic hybrid cell—a precursor of somatic hybrid plant. Thus the production of new hybrids without sexual reproduction will constitute one of the greatest potentials to plant breeders to improve the crop variety. A number of crosses including both intra and interspecific ones are not possible by normal pollination due to sexual incompatibility and therefore the hybrids cannot be produced. Such limitations can be overcome by somatic hybridisation. Genetically controlled character like, disease resistance, protein quality, nitrogen fixation and cold tolerance can be transferred from one species to another by way of protoplast fusion technique. Using this technique a cybrid cell can be produced by fusion of the nuclear material of any one partner and the cytoplasm of both parents.

2.14.1 Methods of Protoplast Fusion

There can be two categories of protoplast fusion.

i. **Spontaneous fusion**   The simple physical contact between similar parental protoplasts can spontaneously fuse to form homokaryon. It is a sort of interspecific fusion that has no practical importance.

ii. **Induced fusion**   The protoplasts isolated from two genotypically different plants can be induced to fuse in the presence of fusion-inducing chemicals or mild electric stimulation to form heterokaryons which have great significance in plant genetics.

The isolated plant protoplasts can be induced to fuse by one of the following ways:

a. **Mechanical Fusion**   Using microscope, micromanipulator and perfusion micropipette, the isolated protoplasts are brought into intimate physical contact mechanically so that the protoplasts can fuse to form a hybrid protoplast.

b. **Chemofusion**   There are several chemicals known as chemical fusogens that induce the fusion of protoplasts. e.g. sodium nitrate, polyethylene glycol, calcium ions, polyvinyl alcohol, etc. In the presence of chemical fusogens, the isolated protoplasts adhere to one another leading to agglutination and in fusion of protoplasts.

c. **Electrofusion**   The mild electric stimulation (10 kv/m) provided by two capillary microelectrodes within the protoplast suspension leads to formation of pearl chain arrangement of protoplasts. Subsequent application of high intensity electric impulse (100 kv/m) for few microseconds results in electric breakdown of plasma membrane at the point of contact leading to fusion of protoplasts.

2.14.2 Identification of fusion product—Hybrid

Proper selection of hybrid cells or fusion products after fusion treatment is necessary since the protoplast populations consist of a heterogeneous mixture of unfused parent protoplasts, homokaryons, and heterokaryons. The microscopic identification of fusion product is based on differences between the parental cell with respect to pigmentation, presence of chloroplast, nuclear staining, cytoplasmic markers, etc. In addition to this, nontoxic flourochromes such as flourescin isothiocyanate or rhodamine isothiocyanate are often used for identification of heterokaryon.

2.14.3 Selection of Hybrids

Some type of selection strategies are necessary at the level of culture to recover the hybrid cells and their callus tissue once the fusion is achieved. Since the cultural behaviour of protoplasts and their nutrient and hormone requirements may vary from plant to plant, several selection procedures have been adopted some of them are as follows:

a. **Biochemical selection**   This selection procedure is based on a prior knowledge of the differential growth characteristics and nutritional requirement of unfused and hybrid protoplasts.

b. **Complementation selection**   The selection of somatic hybrids as a result of complementation by auxotrophic mutants may be used since only the hybrid lines are expected to survive in minimal medium. Auxotrophic mutants require specific compounds for their growth in the culture medium.

c. **Visual selection**   This procedure involves fusion of chlorophyll-deficient (nongreen) protoplasts of one parent with the green protoplasts of other parent (wild type) as it helps in confirmation of heterokaryons under a light microscope.

d. **Morphological selection**   This method of selection of hybrids is based on abnormal morphology. The 'topatoes' are the fusion products of protoplasts of tomato and potato. The regenerated plants showed abnormal morphology and proved to be somatic hybrids by analysis of chromosomal and enzyme fractions.

2.15 CYBRIDS OR CYTOPLASMIC HYBRIDS

These are cells or plants having nucleus of one species but cytoplasm from both the parents. Sometimes the nuclei of two different protoplasts fail to live together in the same protoplast after fusion, eventual loss of the nucleus of one parent plant from the heterokaryon in subsequent stages of development, produces cybrid or cytoplasmic hybrid.

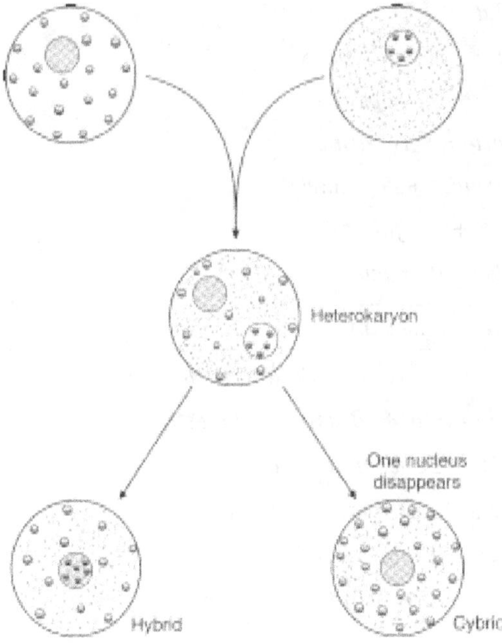

Figure 2.15   Products of fusion of two different protoplasts

Cybrids can be produced in relatively high frequency either by irradiating the protoplasts of one species with x-ray or gamma rays prior to fusion in order to inactivate their nuclei, or by developing enucleated protoplasts (cytoplasts) of one species and fusing them with normal protoplasts of the other species. Cybridisation is one of the ways by which partial genome transfer can be effected in higher plants. Some agronomically important characters such as disease resistance, herbicide resistance and male sterility are controlled by cytoplasmic genes, which are present in either mitochondria or chloroplast. Controlled and directional transfer of only cytoplasmic traits without interference from the nucleus of one parent has been successfully achieved by cybridisation technique.

Table 2.1   Somatic hybrid and cybrid plants generated through protoplast fusion technique

Interspecific Hybridisation
Datura innoxia + D. candida
Lotus corniculatus + L. coimbrensis
Daucus carota + D. capillifolius
Brassica napus + B. compestris
Nicotiana tobacum + N. rustica
Solanum tuberosum + S. phureja
Medicago sativa + M. falcata
Putenia hybrida + P. inflata

Intergeneric hybridisation
Lycopersicon esculentum + Solanum tuberosum
Arabidopsis thaliana + Brassica compestris
Citrus sinensis + Poncirus trifoliate
Datura innoxia + Atropa belladonna
Brassica napus + Eruca sativa
Oryza sativa + Echinochloa oryzicola

The products of protoplast fusion either to produce somatic hybrids or cytoplasmic hybrids (cybrids) are diagrammatically shown in figure 2.15 and some interspecific and intergeneric somatic hybrids are given in table 2.1.

> ## Key Concepts
>
> - Micropropagation is a tissue culture technique used for rapid vegetative multiplication of ornamental plants and fruit trees in a short period of time and in a small place. The rare and endangered plant species are saved by this technique.
>
> - Somatic embryogenesis (SE) is an alternative pathway of regenerating the whole plant by culturing the somatic cell with or without callus proliferation. SE has significant role in production of artificial seeds that have several practical applications.
>
> - Protoplast is naked cell obtained by digesting the cell wall by enzymatic action or by mechanical means. Varieties of plant tissues can be used as sources for isolation of protoplasts. Once the isolation of protoplasts is achieved, its viability should be assessed before the fusion.
>
> - The fusion of protoplasts can be spontaneous or induced by chemical fusogen or electric stimuli. Protoplast fusion technique can generate homokaryon or heterokaryon, of which heterokaryons have immense importance in plant genetics; they can be somatic hybrids or cybrids.
>
> - Cybrids are cytoplasmic hybrids formed by the fusion of two protoplasts where the nucleus of one species and cytoplasm of both parent plants are shared.
>
> - By this technique, it is possible to transfer the genes for disease resistance, drought resistance, rapid growth rate, nitrogen fixation, protein quality, etc. thus producing genetically improved varieties of plants.

Listed below are the various applications of somatic hybridisation and cybridisation.

1. The genomes of two different parents not able to reproduce by sexual means, can be recombined through protoplast fusion technique and a somatic fertile hybrid can be formed.

2. Somatic hybridisation overcomes the barrier of sexual incompatibility.

3. It helps in recovery of recombinants between the parental chloroplast or mitochondrial genomes.

4. Protoplasts of sexually sterile plants can be fused to produce fertile diploids and polyploids.

5. It provides an opportunity to transfer cytoplasmic genes of one species to another species.

Review Questions

1. Define tissue, organ and cell culture.

2. What is culture medium? Describe the types of culture media with their constituents.

3. Define 'Primary culture'. Explain the types of primary culture.

4. What is cell line? Describe the origin and characteristic features of continuous cell line.

5. Briefly explain the process of genetic engineering in the creation of Dolly, the sheep.

6. What are the advantages of animal cloning?

7. Briefly describe the techniques of organ culture and their applications.

8. Write short notes on:

   a. Serum-free media

   b. Clonal propagation

   c. Artificial skin

   d. Cartilage replacement or regeneration

9. What is cell fusion? What are the advantages of cell fusion technique?

10. Explain the experimental procedure for the reactivation of the nucleus.

11. What are monoclonal antibodies? Explain the process of production of monoclonal antibodies.

12. Describe the role of somatic hybridisation in chromosome mapping.

13. Write short notes on:

   a. Sendai virus

   b. HAT medium

   c. Monoclonal antibody

   d. HeLa cells

   e. Chromosome mapping

   f. Nucleus transplant

   g. Hybridoma technology

14. Describe the process of micropropagation with its advantages.

15. Define somatic embryogenesis. How it is useful in the production of artificial seeds? Enlist the practical applications of artificial seeds.

16. Define somatic hybridisation. Explain the process of protoplast fusion.

17. What is protoplast ? Describe the method for isolation of protoplast.

18. What is cybrid? How does this phenomena occur, and does it have any significance in the breeding of plants?

19. Write short notes on:

   a. Sources of plant tissues for protoplast

   b. Cell wall degrading enzymes

   c. Chemical fusogens

   d. Electrofusion

   e. Sexual incompatibility

   f. Cytoplasmic genes

20. Enlist any ten somatic hybrids and cybrids.

3

GENETIC ENGINEERING

3.1 INTRODUCTION

The cell 'world' may be broadly divided into prokaryotic (the cells of bacteria, mycoplasmas and blue green algae) and eukaryotic (cells of protozoa and that making up the tissues and organs of other higher organisms, egg cells, sperms, etc.). The cell has an intrinsic property of reproduction and development due to the presence of 'gene'. A **gene** is a unit of heredity a part of chromosome made up of DNA (deoxyribonucleic acid) or RNA (ribonucleic acid in some viruses). Every gene is responsible for some character or trait of an organism/cell. This is achieved by producing a specific protein through the transcription and translation process with the help of mRNA (messenger RNA).

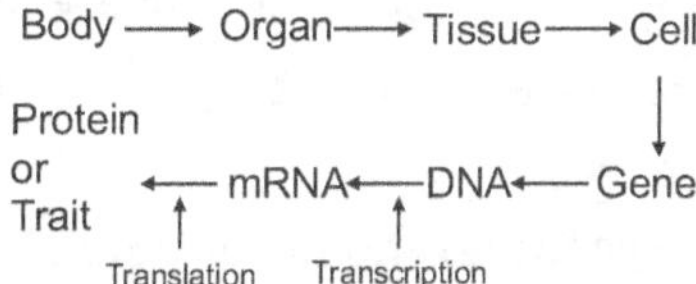

Genetics deals with the study of genes, which has a major share in development and success of biotechnology. Better health, ample food derived from plants and animals, production of useful compounds, low-cost efficient energy, and clean environment, for all this and more, genetics has played a significant role in the achievements in biotechnology which involves applications of scientific and engineering principles to biological processes to provide goods and services. In the last 30 years there has been a tremendous progress in the field of genetics—it has become a technology.

Genetic engineering is defined as the manipulation of genes, using *in vitro* process. These changes may occur as a result of the transfer of a gene from its normal location into a cell, which does not normally contain it, or a gene may have its sequence altered in some way, so that it is a different gene. Such an altered gene is transferred to a new cell type. Thus genetic engineering is nothing but a process of gene manipulation. According to one definition 'Genetic engineering' is also defined as the introduction of manipulated genetic material into a cell in such a way as to replicate and be passed onto the next generation.

Genetic engineering developed in mid-1970s when it became possible to cut DNA at specific sites with the help of **restriction**

enzymes and to transfer particular pieces of DNA containing specific bit of information (a desired gene), from one type of organism into another. As a result, the characteristics of the second organism (recipient) could be changed in a specific way . If the recipient organism is a microbe, such as a single-celled bacterium, the specific fragment of transferred DNA is multiplied many times as the recipient microbe multiplies. Millions of identical cells, i.e. cell clones eventually arise. Consequently it is possible to obtain millions of copies of a specific region of DNA inside a bacterial cell by allowing the cell (and the desired gene) to multiply millions of times. Therefore genetic engineering is alternately called **recombinant DNA technology** (rDNA technology) or **gene cloning**.

The main steps involved in gene cloning are as follows:

1. Isolation of the gene of interest to be cloned using restriction enzymes.

2. Insertion of the gene into another piece of DNA called a cloning vector (it may be a plasmid, phage or virus) which will allow it to be taken up by bacteria and replicate within them as the cells grow and divide.

3. Transfer of the recombinant vectors into bacterial cells (hosts) either by transformation or by infection using viruses.

4. Selection of transformed (transgenic) cells, which contain the desired recombinant vectors.

5. Growth of the bacteria, which can be continued indefinitely, to give as much cloned DNA as needed.

6. Expression of the gene to obtain the desired product, which may be a polypeptide, enzyme, hormone, vaccine, etc. used for human welfare.

In short, gene cloning or rDNA technology is essentially the insertion of a specific piece of 'foreign DNA' into a cell in such a way that the inserted DNA is replicated and handed on to the progeny; and it has important applications in gene mapping, detection of inherited diseases, cancer research, immunology, enzymology, biochemistry and industrial productions.

3.2 RESTRICTION ENZYMES

Enzymes that cut the phosphodiester bonds of polynucleotide chains are called **nuclease**. Those nucleases that preferably break internal bonds are known as **endonuclease.** During the 1970s, it was found that bacteria contained nucleases that would recognise short nucleotide sequences with duplex DNA and cut the phosphodiester backbone at highly specific sites on both strands of duplex and these enzymes are called **restriction endonuclease** or simply **restriction enzymes.** This name was given because they are used by the bacteria to destroy various viral DNAs that might enter the cell, thereby restricting the potential growth of the virus. Thus restriction enzymes serve as a defence mechanism. The bacteria protect their own DNA from nucleolytic attack by methylating the bases at susceptible sites, a chemical modification that blocks the action of the enzyme. Thus restriction enzymes are the molecular scissors that are used to recognise and cut DNA at specific sequences. The sites recognised by them are called **recognition sequences** or **recognition sites**. Different restriction enzymes found in different organisms recognise different nucleotide sequences and therefore cut DNA at different cleavage sites. Table 3.1 lists some restriction endonucleases and the sites at which they cleave DNA.

Table 3.1   Source and recognition sequences (indicated by arrow) of various restriction enzymes

Restriction Enzyme	Source (Organism and strain)	Recognition sequence
Alu I	*Arthrobacter luteus*	↓ A G C T T C G A ↑
Bam HI	*Bacillus amyloliquefaciens H*	↓ G G A T C C C C T A G G ↑
Bal I	*Brevibacterium albidum*	↓ T G G C C A A C C G G T ↑

Table 3.1   *Contd...*

Restriction Enzyme	Source (Organism and strain)	Recognition sequence
*Eco*RI	*Escherichia coli* Ry13	↓ G A A T T C C T T A A G ↑
Hae III	*Haemophilus aegyptius*	↓ G G C C C C G G ↑
Hind III	*H. influenzae* Rd	↓ G T Py Pu A C C A Pu Py T G ↑
Msp I	*Moraxell* sp.	↓ C C G G G G C C ↑
Pst I	*Providencia stuartii*	↓ C T G C A G G A C G T C ↑
Sal I	*Streptomyces albus* G	↓ G T C G A C C A G C T G ↑
Taq I	*Thermus aquaticus*	↓ T C T A A G C T ↑
Xer III	*Xanthomonas oryzae*	↓ C G A T C G G C T T G C ↑

* Pu- any one of two purines (adenine or guanine)

* Py- any one of two pyrimidines (thymine or cytosine)

3.2.1 Recognition Sequences

The sequences recognised by restriction enzymes are 4 to 8 nucleotides long and characterised by a particular type of internal symmetry. Consider the particular sequence recognised by the enzyme EcoRI (figure 3.1).

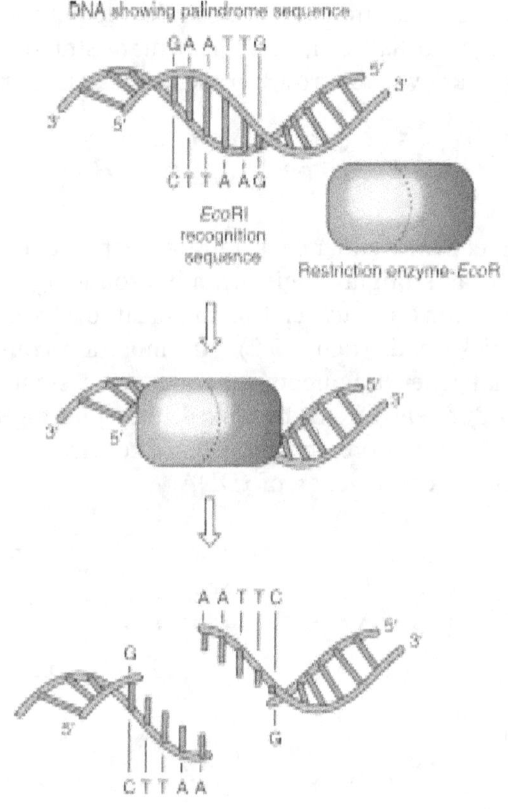

Figure 3.1   Mode of action of EcoRI to cleave DNA

$$5'-G\overset{\downarrow}{A}ATTC-3'$$
$$3'-CTTA\underset{\uparrow}{A}G-5'$$

This segment is said to have two-fold rotational symmetry because it can be rotated 180^0 without change in base sequence. Thus, if one reads the sequence in the same direction (3' to 5' or 5' to 3') on either strand, the same order of bases is observed. A sequence with this type of symmetry is called a **palindrome** when the enzyme EcoRI attacks this palindrome, it breaks each strand at the same site in the sequence, which is indicated by the arrow between the A and G residues.

Thus in a palindrome with rotational symmetry, the base sequence in the first half of one strand of a DNA double helix is the mirror image of the second half of its complementary strand as shown in the following figure where arrow represents the axis of symmetry.

$$5' - G\,A\,A \uparrow T\,T\,C - 3'$$
$$3' - C\,T\,T \mid A\,A\,G - 3'$$

Restriction enzymes either cut (a) straight across the DNA to give blunt ends (b) as straight single strands producing short, single stranded projections at each end to the cleaved DNA to produce cohesive or sticky ends (figure 3.2). An important consequence of this fact is that when fragments generated by a single restriction enzyme from different DNAs are mixed, they join together due to their sticky ends. Therefore, this property of restriction enzyme is of great value for the construction of rDNA.

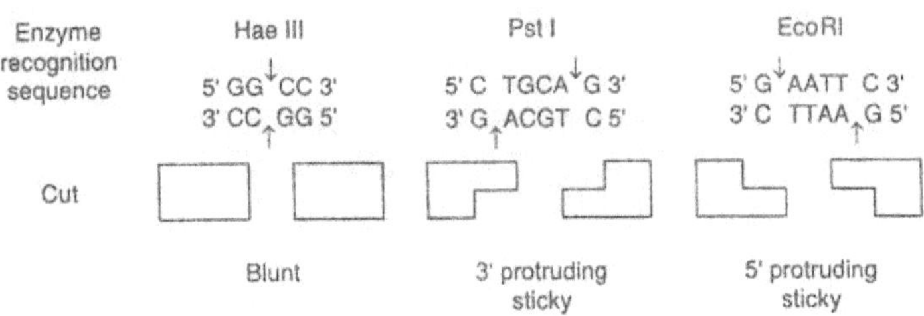

Figure 3.2   Restriction enzymes as molecular scissors

Smith, Nathan and Arber, for their spectacular discovery of restriction enzymes, were awarded Nobel Prize for physiology and medicine in 1978. Restriction enzymes have been used for genetic manipulation by dissecting, analysing and re-configuration of the genetic information at molecular level.

3.3 ISOLATION OF DESIRED GENE

The isolation of a desired gene is the first critical step in gene cloning. The identification and isolation of a gene of interest or DNA fragment called **DNA insert** can be obtained by one of the following ways.

1. cDNA library

2. Genomic library

3. Chemical or enzymatic synthesis of oligonucleotides and

4. Amplification of gene through polymerase chain reaction (PCR)

3.3.1 cDNA Library

cDNA is the copy or complementary DNA produced by using mRNA as a template; while cDNA library is the collection of all such complementary DNA, which is to be used for cDNA insertion in a plasmid or a phage vector.

In this method, mRNA molecule is used to produce cDNA. In 1970 an important discovery was made by Teamin, Mitzutani and Baltimore in relation to reverse transcription, i.e. single-stranded mRNA be converted into double-stranded DNA using an enzyme called **reverse transcriptase** or **RNA dependent DNA polymerase.**

If eukaryotic gene is to be cloned and expressed in prokaryotic cell e.g. a bacterium, then directly cutting the source DNA into suitable fragments alone will not be sufficient because there is difference in gene organisation of eukaryotes and prokaryotes. Eukaryotic genes have introns or noncoding sequences which are transcribed into precursor mRNA then such precursor mRNA undergo post-transcriptional changes involving splicing of introns, addition of poly A (adenylic acid) tail at 3' end and addition of 7-methyl guanosine cap at 5' end. Thus to get expression in the form of protein product after translation of mRNA, the introns have to be removed. Bacteria or yeasts do not have necessary splicing mechanisms for removal of introns. Hence if eukaryotic gene is directly cloned in bacteria or yeast, it will give rise only to precursor mRNA and not the product at the end. This difficulty can be overcome by the cDNA route.

> ### Key Concepts
>
> □ *In vitro* manipulation of genes is called genetic engineering.
>
> □ Millions of identical copies of desired genes obtained with the help of suitable host constitutes recombinant DNA technology or gene cloning.
>
> □ Restriction enzymes are the molecular scissors that are used to recognise and cut DNA at specific sequences called recognition sites and are indispensable for DNA cloning and sequencing.
>
> □ Isolation of a desired gene is the first step in gene cloning and can be obtained by one of the following ways. (i) cDNA library (ii) Genomic library (iii) Enzymatic synthesis of oligonucleotides and (iv) Amplification of gene through PCR.
>
> □ cDNA library is a collection of complementary DNA (cDNA) molecules produced by using mRNA as a template.
>
> □ For making cDNA, a mature mRNA without introns can be directly transcribed into single-stranded DNA with aid of an enzyme reverse transcriptase. Single-stranded cDNA formed is then converted into double-stranded DNA with the help of reverse transcriptase.
>
> □ Use of cDNA is absolutely necessary when the expression of a eukaryotic gene is required in a prokaryote (e.g. a bacterium).

In making cDNA, a mature mRNA without introns can be directly transcribed into DNA using an enzyme reverse transcriptase. cDNA thus produced for a desired protein product can be joined to appropriate vector and cloned into a host. The method is based on the principle that mRNA forms a complex with complementary DNA segment from which it has been transcribed. When eukaryotic mRNA is used as template, a poly–T oligonucleotide (more specifically, oligodeoxynucleotide) is conveniently used as primer since these mRNA have a poly-A tail at their 3' ends. The appropriate oligo-T primer is annealed with mRNA (figure 3.3). Reverse transcriptase extends the 3' end of primer using mRNA molecule as a template.

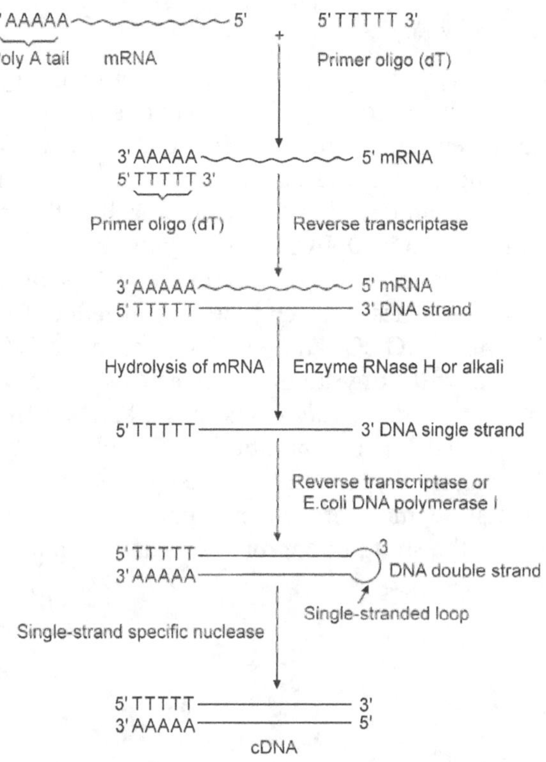

Figure 3.3   Process of formation of cDNA using mRNA as template with the help of enzyme reverse transcriptase

This produces a RNA–DNA hybrid molecule; the DNA strand of this hybrid is obviously the DNA copy (cDNA) of the mRNA strand. Then the RNA strand is digested either by RNase H or alkali hydrolysis, this frees the single stranded cDNA. At the end, cDNA serves as its own primer and provides the free 3'-OH required for the synthesis of its complementary strand; therefore, a primer is not required for this step. The complementary strand of cDNA single strand is synthesised by either the reverse transcriptase or by *E. coli* DNA polymerase I. Since the 3' end of cDNA single strand is used as the primer for this reaction, a short hairpin loop is generated at this end. The hairpin loop is cleaved by a single strand specific nuclease to yield a cDNA.

3.3.2 Genomic Library

One of the most useful reasons for cloning genomic fragments is to produce DNA libraries, which are collections of DNA fragments representing the entire genome of an organism. Genomic library or gene bank includes an entire genome of individual animal, plant, bacteria, and virus under study. In one approach to the formation of a DNA library, genomic DNA is treated with one or two restriction enzymes that recognise very short nucleotide sequences. Two commonly used enzymes that recognise tetranucleotide sequences are *Hae* III (recognises GGCC) and Sau 3A (recognises GATC). Once the DNA is partially digested, the digest is fractionated by gel electrophoresis, and those fragments of suitable size (e.g. 20kb in length) are selected for incorporation into lambda phase particles and used to generate the half million or so plaques needed to ensure that every single segment of a mammalian genome is represented. This constitutes the **shotgun approach** to gene cloning. (figure 3.4).

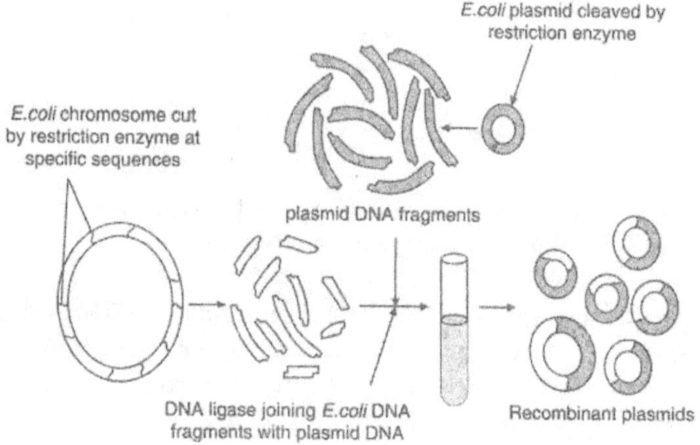

Figure 3.4   Method of formation of genomic library using rDNA technique

The following are the important steps in the construction of genomic library:

i. Chromosomal DNA of interest is isolated.

ii. The fragments of suitable size are cut either randomly by shearing or with suitable restriction enzymes.

iii. The desired fragments of DNA are cloned in suitable vector.

   iv. Clones are screened using gene probe, then identified and characterised.

   v. Amplification of the clones is carried out on culture medium to maintain the genomic library.

The number of clones or bacterial colonies of a genomic library depends on the following factors.

1. The complexity of the genome (the more complex the genome the larger the size).

2. The DNA insert or fragment length used for cloning (the smaller the fragment size, the larger the number of clones for the same genome). The minimum number of clones in a genomic library constructed with 20 kb inserts for a 99% probability of all genomic sequences being represented would be 1,157 for *E. coli* (genome size 4.6×10^6 bp), 3,462 for *Drosophila* (genome size 180.0×10^6 bp) and 6,90819 for man (genome size 3200×10^6 bp).

3.3.3 Chemical or Enzymatic Synthesis of Oligonucleotides

The development of chemical techniques to synthesise polynucleotides having a specific base sequence was begun by H.G. Khorana in the early 1960s as a part of an attempt to decipher the genetic code. Khorana and his co-workers continued to elaborate their techniques, and in 1970, they succeeded in synthesising a complete bacterial tyrosin tRNA gene, including the non-transcribed promoter region. The gene, totalling 126 base pairs, was put together from over 20 segments, each of which was individually synthesised and later joined enzymatically. This artificial gene was then introduced into bacterial cells carrying mutations for this tRNA, and the synthetic DNA was able to replace the previously deficient function.

A second landmark in the area of gene synthesis came in 1977 when a gene coding for the small hypothalamic peptide hormone, somatostatin (14 amino acid residues) was synthesised by Keiichi Itakura and his co-workers of the city of Hope Medical Center. The gene was inserted into a specially constructed plasmid downstream from bacterial regulatory sequences and introduced into *E. coli*, where it was transcribed and translated. A gene for the first 'average sized' protein, human interferon, was synthesised in 1987, an effort that required the synthesis and assembly of 67 different fragments to produce a single DNA duplex of 514 base pairs containing initiation and termination codons recognised by the bacterial RNA polymerase to transcribe and translate this synthetic gene.

In enzymatic synthesis of DNA, the bacterial enzyme polynucleotide phosphorylase is used. Though the enzyme is specific for ribonucleotides it polymerises deoxyribonculeotides at a slow rate when Mn^{2+} replaces Mg^{2+} in the reaction mixture. The process requires a primer of at least 3 nucleotides long. The enzyme adds one or two nucleotides at a time to the chain and thus the chain continues to grow in the 5'→3' direction as shown in the figure 3.5.

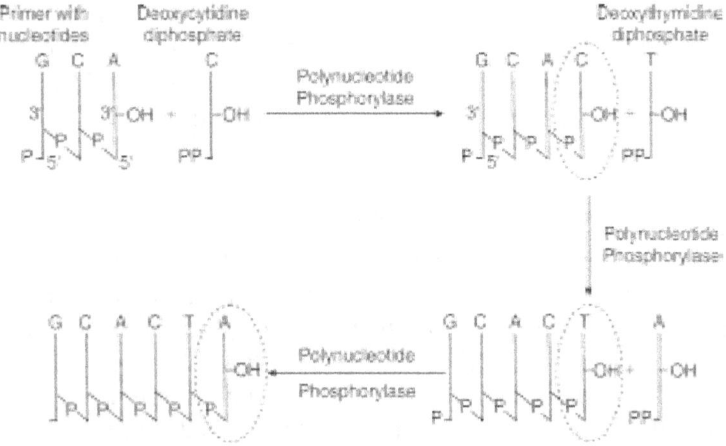

Figure 3.5   Enzymatic synthesis of oligonucleotide

In the last five years, the development of research facilities in the field of molecular biology has improved so much that automated DNA synthesising machines are commercially available. These machines are able to synthesise polynucleotides of any desired sequence approximately 50 nucleotides in length. In this approach, silica-based or controlled-pore glass beads solid support is used for automated synthesis of oligonucleotides. It takes less than 15 minutes for adding mononucleotide to the chain. The automated DNA synthesisers are popularly called gene machines; they are microprocessor-controlled and carry out all the operations automatically (figure 3.6). Once such fragments are synthesised, they can be covalently joined to one another to generate synthetic DNA molecules of considerable length.

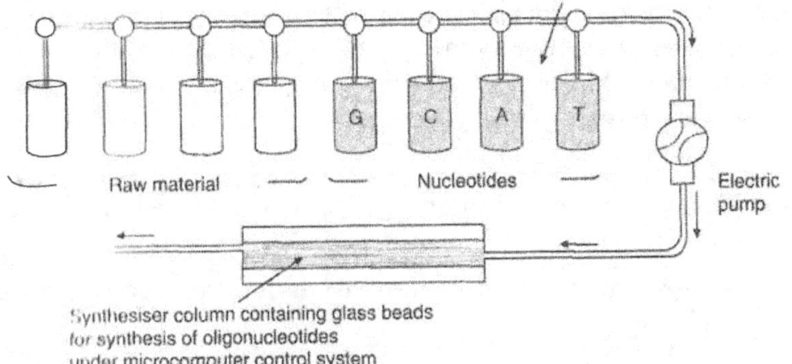

Figure 3.6   Automated gene synthesiser

Chemically synthesised oligonucleotides have several applications in gene biotechnology.

1. Oligonucleotides are used as DNA hybridisation probes for gene isolation and in diagnosis of diseases.

2. They are used as primers for DNA sequencing.

3. They assist in altering the sequence of cloned genes.

4. They help in the amplification of specific DNA segments using PCR.

5. Oligonucleotides are also used as linkers and adaptors in gene cloning.

6. They serve as therapeutic tools for viral inhibition or cancer therapy and also help in studying oncogene activation.

7. They are used to produce complete gene sequences by linking them in a defined order.

8. They also aid in *in situ* hybridisation to study gene expression in animal or human tissue.

3.3.4 Amplification of Gene through Polymerase Chain Reaction (PCR)

In 1985 Kary Mullis made an important discovery in the form of an extremely powerful technique called polymerase chain reaction (PCR).

PCR can generate a billion copies of the desired segment of DNA or RNA, present even as a single copy in the initial preparation with high accuracy and specificity in a matter of few hours. The process of PCR is completely automated and involves automatic thermal cycles for denaturation and renaturation of double stranded DNA. The device required for PCR is called thermal cycler that is commercially available in the market.

PCR is *in vitro* amplification of a desired DNA segment, which involves.

   i. a DNA preparation containing the desired segment to be amplified.

   ii. two complementary nucleotide primers of about 20 bases long.

   iii. four deoxyribonuclueoside triphosphates namely,

      d TTP (deoxythymidine triphosphate),

      d CTP (deoxycytidine triphosphate),

      d ATP (deoxyadenosine triphosphate), and

      d GTP (deoxyguanosine triphosphate).

   iv. a heat stable DNA polymerase.

Heat stable DNA polymerase Earlier DNA polymerase enzyme from *E. coli* was used. But as it is sensitive to heat, which destroyed the enzyme, after each cycle new enzyme had to be added. Use of thermostable DNA polymerase enzyme, for example, Taq polymerase from *Thermophilus aquaticus* bacteria present in hot spring, was reported by Saiki *et al* in 1988 and that was the major breakthrough in PCR development. Other thermostable DNA polymerase enzymes used today in PCR are Pfu polymerase isolated from *Pyrococcus furious* and vent polymerase from *Thermococcus litoralis*. Both these polymerase enzymes are more efficient than Taq polymerase.

Mechanism of PCR At the start of PCR, the DNA segment, an excess of two primer molecules, four deoxyribonculeoside triphosphates and the thermostable DNA polymerase are mixed together in 'eppendorf tube' and the following operations are performed sequentially (figure 3.7).

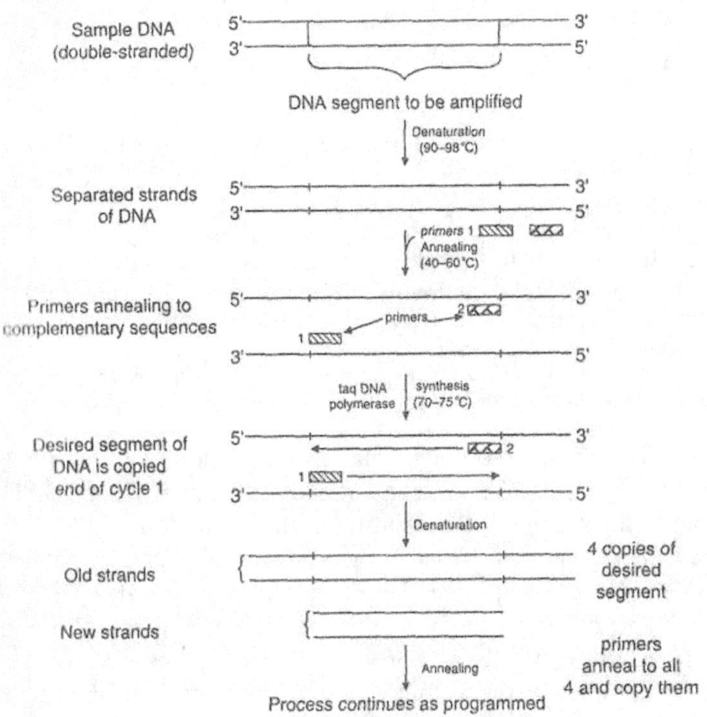

Figure 3.7   Steps involved in polymerase chain reaction (PCR)

Step 1   The reaction mixture is heated to a temperature (usually 90–98 °C) to separate two strands of desired DNA. This is called denaturation.

Step 2   The mixture is allowed to cool at a temperature (generally 40–60 °C) that permits annealing of the primer to the complementary sequences in DNA; these sequences are located at the 3' end of two strands of desired DNA segment and this step is called annealing.

Step 3   The temperature (usually 70–75 °C) allows thermostable Taq DNA polymerase to use single-stranded DNA as template and by utilising primers for both the strands of DNA new complementary strands are synthesised. This step is called renaturation.

In an automatic thermal cycler, the above three steps are automatically repeated 20–30 times as predetermined by the computer. The programme in thermal cycler is set to denature at 94 °C for 20 sec, anneal at 55 °C for 20 sec., extend and renature at 72 °C for 30 sec.

Thus the completion of the above 3 steps completes the first cycle of amplification and each cycle takes 3.75 min. At the end of the first cycle, two identical copies of DNA are formed.

Step 4   It begins the next cycle of amplification of freshly synthesised double stranded DNA. It involves denaturation, i.e. separation of DNA strands similar to the step 1.

Step 5   Annealing allows the primers to base pair with separated strands of DNA.

Step 6   Synthesis of complementary strands, doubling the number of copies of desired DNA segment present at end of step 3. This completes the second cycle resulting in the formation of 4 copies of desired DNA segment. Thus at the end of 'n' cycles 2^n copies of DNA segment are amplified. The researcher has to only specify the number and duration of cycles, the machine performs the entire operations automatically and precisely. Once the desired number of cycles is completed, the amplified DNA segment is purified by gel electrophoresis. After its sequencing, the amplified DNA segment can be inserted into a traditional cloning vector if expression is desired.

Applications of PCR   PCR has several novel applications. Some of them are as follows:

1. PCR helps in the detection of pathogens in food, water and other samples though the number of organisms is very low.

2. Detection of specific genes for the pathogenicity of bacteria is possible by PCR. Dr. V. Sritharan from Anna University, Tamilnadu has developed an instant diagnostic test for tuberculosis.

3. Detection of AIDS virus itself (direct method) instead of detection of antibody of HIV (indirect method) can be done with PCR.

4. PCR helps in prenatal diagnosis of genetic diseases like sickle cell anaemia, phenylketonuria, β thalassemia, haemophilia, etc.

Key Concepts

- Genomic library is the collection of entire genome of an individual animal, plant, bacteria or virus under study.

- Gene bank is constructed by isolating the complete genomic DNA from cell.

- H.G. Khorana and his co-workers were the first to succeed in synthesising a complete bacterial tyrosin tRNA gene including the non-transcribed promoter region.

- With advancement in this research, Automated DNA synthesising machines are now available. Chemically synthesised oligonucleotides have immense importance in (i) disease diagnosis (ii) cancer therapy (iii) study of oncogene activation (iv) *in situ* hybridisation to study gene expression (v) amplification of specific DNA segments (vi) gene cloning as linkers and adaptors.

- Polymerase Chain Reaction (PCR) is an extremely powerful technique, which can generate a billion copies of desired DNA or RNA segment with high accuracy and specificity.

5. It is possible to determine the sex of human and livestock embryos, fertilised *in vitro* before implantation with help of PCR.

6. The problems in identification of criminals, disputed parentage can be tackled effectively by DNA fingerprinting through PCR from a piece of hair, blood stain or seminal fluid, etc. having partially degraded DNA. Characterisation of an individual is possible through DNA amplification via PCR. In India recently the problem related to Tandoor case of Naina Sahani, parentage dispute of child related to Amarmani Tripathi and Madhumita murder case were solved by this technique only. Even the confirmation of the identity of Saddam Hussein in Iraq was done by gene amplification through PCR.

7. It is also possible to study the genetic profile of animals like dinosaurs, mammoth and human beings, which lived few thousand years earlier with the use of PCR.

8. Genetic and physical mapping of chromosomes of plants and animals can be prepared using PCR.

9. One of the significant uses of PCR is to study DNA polymorphism in genome using known sequences as primers. Synthetic nucleotides of any sequence can be used as random primers to amplify polymorphic DNAs having sequences specific to the primers used. Such an application of PCR generates Random Amplified Polymorphic DNA (RAPD) which is detected as bands after electrophoresis. It has been used for study of chromosomal mapping in maize, soyabean, mouse, man, etc.

3.4 VECTORS

Vectors are the DNA molecules having the ability to self-replicate and into which foreign DNAs or genes of interest are integrated to form recombinant DNAs. Then they are introduced into appropriate hosts for gene cloning or expression of the gene of interest. Vectors can be any extra-chromosomal small genome. A plasmid, phage, cosmid or virus may serve as a vector.

3.4.1 Types of Vectors

There can be two types of vectors, namely cloning vector and expression vector.

Cloning Vector   These vectors are used for propagation of DNA inserts in a suitable host. There is no transcription or translation of cloned gene in such vectors. Cloning vectors are useful in the creation of a genomic library or Gene bank, in the preparation of molecular probes or in genetic engineering experiments.

Expression Vector   These vectors are designed to express the cloned gene i.e. they allow the transcription and translation of the cloned gene to produce a protein. This can be achieved by adopting certain strategies if a eukaryotic gene is to be expressed in a prokaryote. Expression vectors are used in producing transgenic plants or animals so that the cloned gene expresses itself to give products (protein) that can be utilised for human welfare.

3.4.2 Properties of a Good Vector

A good vector should have the following properties.

i. It should have the ability of independent replication so that as the vector replicates and large number of copies of the DNA insert will be formed.

ii. It should be able to easily introduce into host cells, i.e. transformation of host with the vector should be easy.

iii. A vector must contain unique target sites for restriction enzymes into which the DNA insert can be integrated without disrupting the function.

iv. When the expression of DNA insert to be achieved, the vector should have at least suitable control elements like promoter, operator, ribosomal binding sites, etc.

Among all bacteria, *E.coli* is the best host, since it is very easy to clone and isolate the DNA insert in it. In 1973 Cohen *et al* for the first time described the successful construction of a cloning vector, plasmid pSC 101 for use in *E. coli* as host. Today there are more than 1000 fully described cloning vectors. Since *E. coli* is easy to transform and allows the independent replication of a DNA insert, it supports several types of vectors such as (i) plasmids, (ii) bacteriophages, (iii) cosmids and (iv) viruses.

Plasmids   These are extra-chromosomal, small, circular, double-stranded DNA molecules capable of self-replication and present in bacterial cells. Plasmids can integrate themselves with the main bacterial chromosome to form episomes. The independent plasmids are not essential for bacterial cells except under specific environments. There are several types of bacterial plasmids, some of them are given below.

a. F plasmids responsible for bacterial conjugation.

b. R plasmids carrying genes for resistance to antibiotics. and

c. Col plasmids that code for the proteins called colicins that kill sensitive *E.coli*.

Plasmid Vectors   Most of the plasmids, which are used as cloning vectors in *E. coli*, are derived from natural plasmids. The natural plasmids are modified to form cloning vectors. Plasmid DNA used as vectors can be cleaved at a site with restriction enzymes where foreign DNA can be inserted. Some of the reconstructed plasmid cloning vectors are as follows.

a. **pSC 101**   It is the earliest plasmid vector containing the replication module (ori) for replication in *E.coli*, tet^r gene for resistance to tetracycline and single recognition site for restriction endonuclease EcoRI, Hind III, Bam HI, Sal I as shown in figure 3.8.

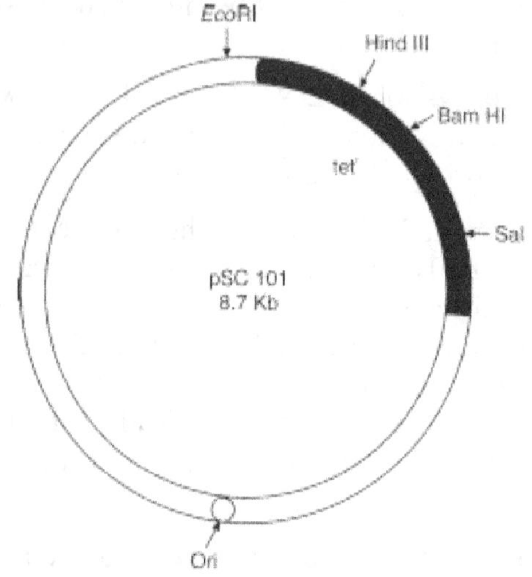

Figure 3.8   Plasmid vector—pSC 101

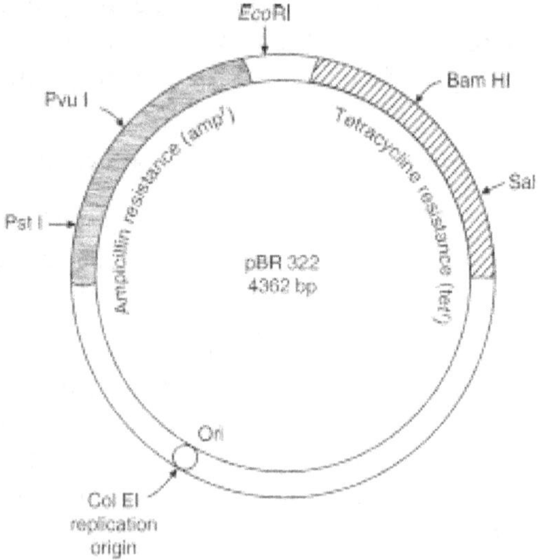

Figure 3.9   Plasmid vector—pBR 322

b. **pBR 322**   Its name has p for plasmid, B and R for Boliver and Rodriguez, the scientists who developed it. Its structural features include Col E1, origin of replication that allow productions of several copies per cell and it has two genes for different antibiotic resistance i.e. ampicillin and tetracycline resistance. (amp^r and tet^r) as shown in figure 3.9.

c. **pUC**   It is the most popular and widely used plasmid. Its name is derived from the place of its initial preparation i.e. University of California. It has 2686 bp, ampicillin resistance gene and LacZ gene. Vectors of pUC family have several restriction sites in a small stretch of DNA, such region is called multiple cloning site (mcs) as shown in figure 3.10.

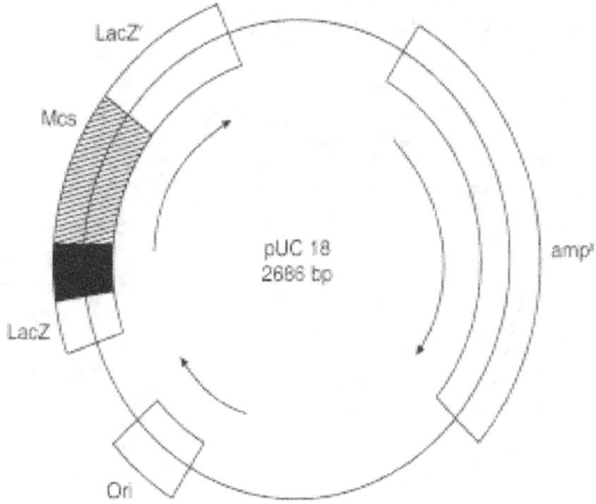

Figure 3.10   Plasmid vector—pUC 18

Bacteriophages as cloning vectors   Bacteriophages are the eaters of bacteria. Most phages lyse the bacteria they infect (lytic phages) while many other phages integrate their chromosome into the main bacterial chromosome and multiply (lysogenic phages). Phage has a linear DNA, so a single break creates two fragments. Foreign DNA can be inserted between them and two fragments can be joined. Such phages when undergoing a lytic cycle in the host will produce many recombinant DNA. λ (lambda) and M13 phages are phages of *E.coli* that are mostly used as cloning vectors. The genome of phage contains 48,502 bp having an origin of replication, genes for different proteins to form head and tail, enzymes and single-stranded protruding cohesive ends. Its genome remains linear in phage head, but when it infects *E.coli*, its cohesive ends anneal to form circular molecule essential for replication. The joined cohesive ends are called cos sites with 12 bases in length. Replication of DNA by rolling circle mechanism results in a concatemer which is a long molecule made up of many copies of viral DNA lined end to end through cos sites.

Cosmids as cloning vectors   These are cloning vectors having the properties of both plasmid and λ phage. They are essentially plasmids having a minimum of 250 bp of λ phage DNA. A typical cosmid has a replication origin, unique restriction site, selectable markers from plasmid as shown in figure 3.11.

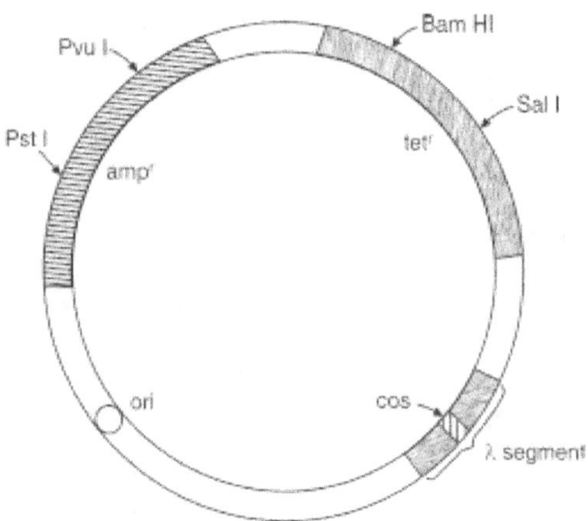

Figure 3.11   Structural features of cosmid vector

Cosmid vector is made up of pBR 322 and λ segment with cos sites, a Pst I and Pvu I recognition site for restriction enzymes within ampicillin resistance gene and *Bam* HI and *Sal* I sites within tetracycline resistance gene.

Barbara Hohn and John Collins developed cosmids in 1978. The use of cosmid vectors for cloning of DNA insert is advantageous because cosmids can accommodate up to 45 bp long DNA inserts. Their efficiency is high enough to produce a complete genomic library of 10^6–10^7 clones from just 1 µg of DNA insert. Cosmids have been successfully used in producing gene banks of *Drosophila*, mouse and several other organisms.

Viruses as cloning vectors   The self-replicating structures like bacterial plasmids are absent in animal cells hence some viruses are used to develop the vectors for animal cells to be transformed. Animal cells used for transformation may be obtained from tissue culture cell lines, *Xenopus* oocytes, early embryo or even body tissues of individuals aimed either for producing transgenic animals or for gene therapy and vaccination. Virus has two life cycles depending on host cell line employed. Some vectors are capable of replication while others are not, categorising into replicating or non-replicating vectors respectively. Further any segment of DNA, or a mixture of fragments become integrated into cellular genome and express themselves to produce a product. SV 40 (simian virus 40), adenoviruses, papiloma virus and vaccina (provirus) having genome in the form of DNA and retroviruses containing RNA genomes are the examples of animal viruses used as cloning vectors.

3.5 PROBES

The small fragments of DNA or RNA usually 15–30 bases long and used to identify the complementary sequences in nucleic acid samples are called probes. This is possible by allowing the probes to form base pairing with the sample nucleic acids and then identifying the samples that show complementary sequences with that of probes. This is called nucleic acid hybridisation, which is a very precise, rapid, and extremely reliable method provided probes should be suitably labelled for easy detection.

3.5.1 Types of Probes

Molecular probes like genomic DNA, cDNA, synthetic oligonucleotides or RNA probes can be used for nucleic acid hybridisation. These probes differ only in the methods of their preparation but ultimately they serve the same purpose of identification of specific sequences of DNA or RNA in the samples of nucleic acid. DNA probes are more convenient and preferable.

Consider a situation where one has a mixture of hundreds of fragments of DNA of identical length and overall base composition that differ from one another solely in their base sequence. Assume, for example that one of the DNA fragments constitutes a portion of a β-globin gene and the other fragments contain unrelated genes. The only way to distinguish between the fragment containing the information for β-globin polypeptide and all the others is to carry out a molecular hybridisation experiment using complementary molecules as probes.

In the present example, incubation of the mixture of denatured DNA fragments with an excess number of β-globin mRNAs would drive the globin fragments to form double stranded DNA–RNA hybrids, leaving the other DNA fragments single stranded. There are a number of ways to separate the hybrid from the single-stranded fragments.

3.5.2 Applications of DNA Probes

1. They help in the detection of recombinant clones carrying the desired DNA insert.

2. DNA probes are useful in accurate diagnosis of diseases caused by various parasites, pathogens or viruses.

3. Using molecular probes it is possible to detect pathogenic viruses in agricultural produce and animals.

4. They are employed for confirmation of the integration of a DNA insert in the host genome and its expression in transformed cells.

5. They assist in forensic investigations in relation to criminal identification, parentage dispute or establishing family relationship.

6. Genetic disorder in humans may arise due to change in a single base pair. It is possible to detect such diseases by ante-natal diagnosis. DNA probes successfully differentiate RFLP (Restriction Fragment Length Polymorphism) pattern for a normal and diseased or altered gene. Thus prenatal diagnosis of genetic defect in foetal DNA is possible at 6–8 weeks of pregnancy.

3.6 DNA SEQUENCING

The characterisation of DNA is essential to understand the function of gene. Knowing the nucleotide or base sequence of a DNA molecule is called DNA sequencing. On the basis of this knowledge the amino acid sequence in protein and its molecular weight can be understood. The information regarding the detailed analysis of base sequences is useful in molecular cloning studies and in regulation of gene expression. It is possible to detect the nucleotide sequence because of the following advancements.

a. the availability of restriction enzymes.

b. the development of highly sensitive gel electrophoresis technique to separate DNA fragments differing by only one nucleotide.

c. large number of copies of individual DNA fragments can be made available as a result of gene cloning and PCR technique.

There are two common methods used for DNA sequencing.

a. Maxam and Gilbert's chemical degradation process.

b. Sanger's dideoxyribonucleotide synthetic method.

3.6.1 Maxam and Gilbert's Chemical Degradation Process

Allan Maxam and Walter Gilbert of Harvard University developed this method. In this method chemical reagents are used to destroy specific nucleotide bases and thus break the DNA molecule at specific sites. The steps involved in this process are as follows:

1. The strands of the DNA molecule are radioactively labelled by the addition of ^{32}P-dATP at one end (usually the 5' end).

2. Two strands of DNA are separated, as only one will be sequenced.

3. End labelled single-stranded DNA sample is treated with four different chemical reagents that break the strand at one end or at two specific nucleotides. The cleavage is allowed to occur at G, G+A, C+T or T .

4. The digest from the four reaction mixtures are then separately subjected to gel electrophoresis to separate the fragments by their size.

5. Determination of nucleotide sequence by the pattern of radioactive band developed in the four lanes of gel on X-ray film (figure 3.12).

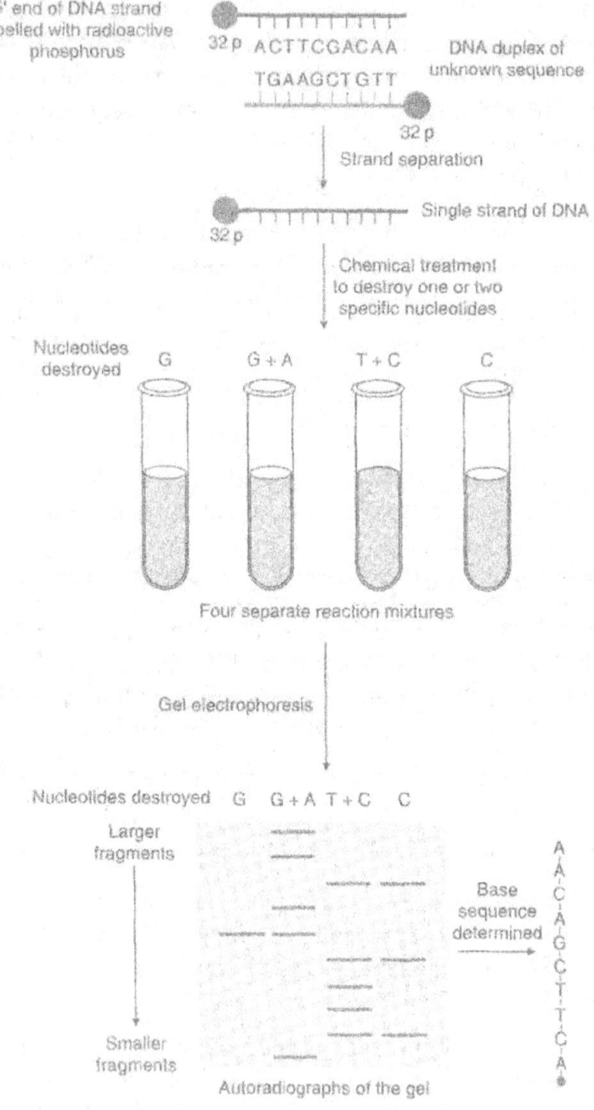

Figure 3.12   DNA sequencing method developed by Maxam and Gilbert

3.6.2 Sanger Dideoxyribonucleotide Synthetic Method

In 1977, Fredrick Sanger and A.R. Coulson of the Medical Research Council in Cambridge, England developed the technique for DNA sequencing. In the normal process for replication, a deoxyribonculeoside triphosphate (dATP, dGTP, dCTP or dTTP) has a 3' hydroxyl group of the sugar moiety to form a phosphodiester bond, while dideoxynucleotide (ddATP, ddGTP, ddCTP and ddTTP) lacks 3' OH group. Therefore during replication, phosphodiester bond cannot form and it will serve for stopping the chain elongation. The involvement of ddNTP in the process and steps in the method are as follows:

1. Begin the process with a population of identical DNA segments to be sequenced up to about 500 base pairs in length that are obtained by treatment of DNA with restriction enzyme.

2. Then divide the preparation into 4 samples, each of which is treated in a slightly different way.

3. The DNA of each sample is denatured into its single strand (such single strand samples of DNA segment can be obtained by cloning them in M13 virus).

4. Incubate single-stranded DNA with a short, radioactively labelled oligonucleotide (acting as primer) that is complementary to the 3' end of one of the single-stranded fragments. The reaction mixture also contains an enzyme DNA polymerase, all four deoxyribonculeotides (dNTP) and a low concentration of a modified precursor, called dideoxyribonucleoside triphosphate (ddNTP). A different ddNTP (ddGTP, ddATP, ddGTP or ddTTP) is added to the four samples.

5. During incubation of the reaction mixture under appropriate conditions, the labelled oligonucleotide binds to the 3' end of single-strand fragments and serves as a primer for the addition of nucleotide into a growing complementary chain when ddNTP is randomly incorporated in the place of normal dNTP causing chain termination. As a result incomplete, radioactively labelled DNA fragments of different lengths are formed.

6. Fragments formed in the four reaction mixtures are loaded into separate wells on gel electrophoresis.

7. The gel electrophoresis is used for autoradiography to visualise a band and from their position the sequence of entire DNA molecule can be read directly (figure 3.13).

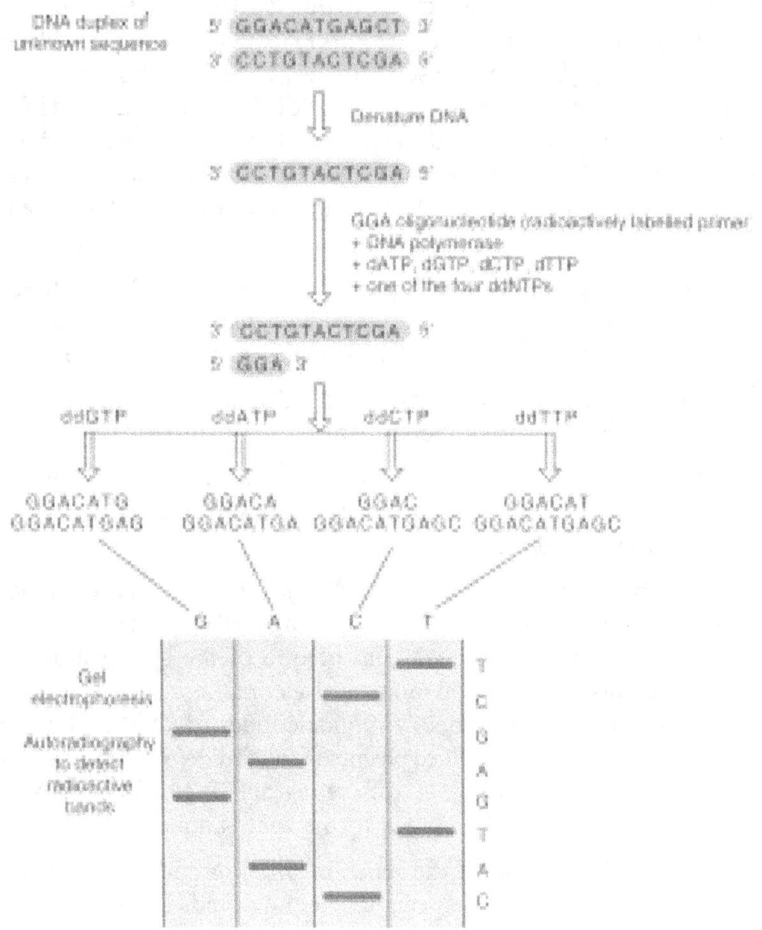

Figure 3.13   Sanger's dideoxyribonucleotide synthetic method for DNA sequencing

Once the nucleotide sequence of DNA segment is determined, it is relatively a simple task to deduce the amino acid sequence of an encoded polypeptide. Once the amino acid sequence has been deduced, it can be compared to other known sequences to provide information about the polypeptide's possible function. Thus sequence of DNA provides the knowledge of every gene, its transcript, the intron and the timing of transcript synthesis. DNA sequencing technology is used in the attempt to determine the nucleotide sequence of the entire human genome.

In1998, the nematode worm *Caenorhabditis elegans* was the first multicellular organism to have its complete genome sequenced. A team lead by Sir John Sulston and Allon Coutson at the Welcome Sanger Institute in Hinxton, Cambridge, successfully completed this project, which proved to sceptics that existing sequencing technology could be scaled up to tackle the human genome. Although this nematode worm has only 19,000 genes (the human genome has around 35,000 genes) the worm is providing to be an invaluable model organism for the study of gene function. Because it has a short lifespan and ages quickly, the worm is also a useful model for studying ageing. And it can provide valuable insights into human diseases such as cancer.

In 2000, the genome sequencing of the common fruitfly *Drosophila* was completed while in December 2000, the publication of the first whole genetic sequence of the plant, *Arabidopsis thaliana* (Thale cress) in *Nature* marked a major breakthrough in our understanding of how plants work and will offer plant biologists and plant breeders significant opportunities for making plants easier to grow and healthier to eat. This project was initiated and led by the John Innes Center in Norwich. Scientists have already started to isolate useful genes. For example the gene that controls flowering time in Arabidopsis. Isolation of this gene, FRIGIDA, opens up opportunities for plant breeders to develop winter and spring varieties of crops.

Research at Horticulture Research International, UK, is using the *Arabidopsis* gene sequence to identify natural disease resistance genes that have been lost from related members of the *Brassica* (cabbage) family as a result of selective breeding. This may make it possible to restore natural resistance to diseases in vegetables such as cabbages, cauliflower, broccoli and Brussels sprout and reduce current reliance on chemical controls.

3.7 GENETIC ENGINEERING FOR COMMERCIAL PRODUCTION OF INSULIN

As stated earlier genes form the fundamental basis of all life, determine the properties of all forms of life and are defined segments of DNA.

Recombinant DNA technique, popularly known as gene cloning or genetic engineering, offers potentially unlimited opportunities for creating new combinations of genes that do not exist under natural conditions. Genetic engineering involves the formation of new combinations of heritable material by the insertion of nucleic acid molecule, produced by whatever means outside the cell, into any virus, bacterial plasmid or other vector system so as to allow incorporation into a host organism (it may be a bacterium, yeast, plant or animal cell) in which they do not naturally occur but are capable of continued multiplication.

Genetic engineering permits the selective alteration of the DNA in bacterial cells so that new genes are added to the DNA of the bacteria, with the result that the altered cells can synthesise nonbacterial protein like insulin, interferon, vaccines, growth hormones and antibiotics which are of immense biomedical or economic importance. It is possible to obtain an unlimited and chemically pure supply of these proteins with the aid of genetically engineered microbes. These techniques are useful not only in the study of basic processes in developmental biology (differentiation) but also in the actual alleviation of disease by gene transfer. In addition to this, rDNA technology is useful in the generation of new strains of crop plants (transgenic plants) and animals (transgenic animals) having altered characteristic features directed towards human welfare.

Recombinant technology has widespread applications, which not only involve the insertion and reproduction of eukaryotic genes in prokaryotic bacteria, but also the insertion, reproduction and expression of recombinant genes in eukaryotic cells and whole organisms. The various steps involved in genetic engineering (as mentioned earlier) are explained in brief with reference to manufacture of human insulin as follows.

3.7.1 Identification and Isolation of Insulin Gene

In the laboratory, the first step towards creating a genetically altered microbe, which will manufacture a human protein, is to identify and isolate the gene from a human cell that codes for insulin. With a few exceptions all cells of an organism contain the same genetic information, but they all do not use that information in the same way. The cells of different organs have different functions to perform, and this specialisation among cells is because of gene expression. This difference in expression between genes in specific cell types is used by genetic engineers as the most elegant method of isolating the required gene. This technique focuses on the cells' mRNA molecule, rather than searching through huge complex mass of genes encoded in DNA. Thus, if a gene to make insulin is required, human pancreas cells are examined to find mRNA molecule, which have been transcribed from that gene.

First the mRNA molecule must be extracted from the pancreatic cells. When these cells are broken open, they release materials including not only mRNAs but also enzymes, other proteins, DNA and pieces of cell membrane. All these materials are separated from each other by treating the mixture with chemicals and subjecting it to a variety of physical forces. By centrifugation at various speeds and timing, different kinds of molecules are separated according to their weights and shapes.

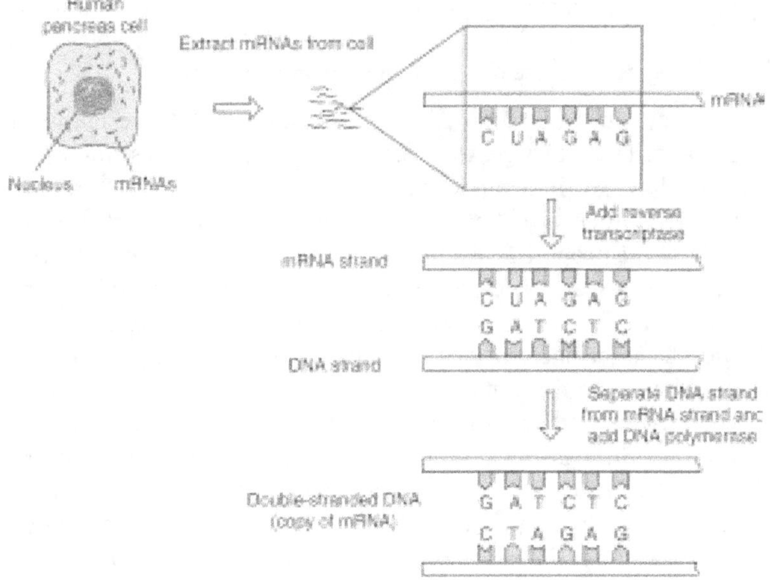

Figure 3.14   Process of formation of cDNA from mRNA by the enzyme reverse transcriptase

The isolated population of mRNA molecules (mixed/ heterogeneous) contains the genetic information needed to manufacture all the proteins in the form of codons. This information is converted back into the form of the equivalent DNA sequences. With the help of a special enzyme, reverse transcriptase (RNA dependent DNA polymerase), it is possible to synthesise new single-stranded DNA molecule. This enzyme assembles the nucleotides in a linear sequence according to the order of bases on the mRNA, thus it uses mRNA as a 'template'. After a single strand DNA molecule has been copied from isolated mRNA, a complementary strand of DNA is produced using the standard DNA copying enzyme, DNA polymerase. Thus a new population of double-stranded 'copy' DNA (cDNA) molecules is produced that codes for mRNA that were in differentiated tissue. In other words, cDNA contains the same genetic information as the mRNA from which it has been formed. This cDNA is a copy of human gene (figure 3.14). Since the process started with many different types of mRNA isolated from human pancreas cell, there are now cDNA copies of many different genes among which are copies of the insulin gene.

3.7.2 Integration of cDNA into the Plasmid (vector)

Plasmids are small, circular extra-chromosomal DNA present inside the bacteria. They carry genes which enable bacteria to resist antibiotics. Plasmids have an independent ability of replication and they often pass from one cell to other even though the cells are of different species. These properties are most important for genetic engineering. Plasmids are used as vectors and are taken from one set of bacteria. The human cDNA gene is stitched into a plasmid circle and this is allowed to enter bacteria and transfer the human gene into its new home. The plasmid is cut open with a restriction enzyme that has a unique target site located in the sequence where the cDNA is to be inserted. *Bam*HI is the restriction enzyme used to split open the plasmid ring, leaving a set of 4 unpaired bases at each end of the molecule as shown in figure 3.15. The same restriction enzyme is used to clear the cDNA but before the action of the restriction enzyme, at each end of the cDNA, six linker bases are added. The restriction enzyme clears the cDNA through the linkers, producing unpaired bases that match those on the plasmid. The cDNA and the plasmid are then brought together to produce a single circular molecule. This is a recombinant DNA molecule or recombinant vector having integrated human gene cDNA. At this stage both the genes (cDNA and plasmid) are held together by 8 complementary weak bonds at their sticky ends. With the help of the enzyme DNA ligase, a permanent link is established between both the ends of genes providing stability to the recombinant plasmid.

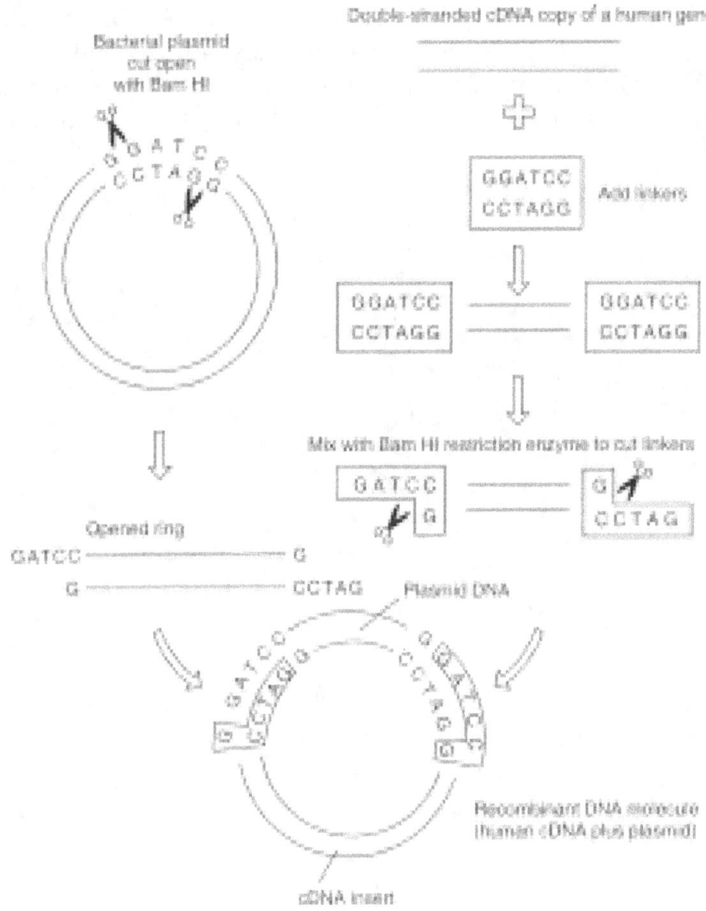

Figure 3.15   Method of preparation of recombinant DNA molecule

3.7.3 Introduction of Recombinant Vector into a Suitable Host

Once *in vitro* construction of recombinant vector is completed, it is the bacteria that are to act as factories manufacturing the desired protein. The most popular choice of bacterial host is *Escherichia coli*. Plasmids have an inherent ability to enter the cells of *E.coli*, and adding a few simple chemicals to the mixture can facilitate this invasion. Once inside a bacterial cell, a single plasmid may multiply itself to produce a few dozen identical replicas. As the bacterium which harbours the recombinant plasmid is also growing and dividing as often as once every 20 minutes, each daughter cell takes with it a few of the plasmids, which again reproduce themselves (figure 3.16).

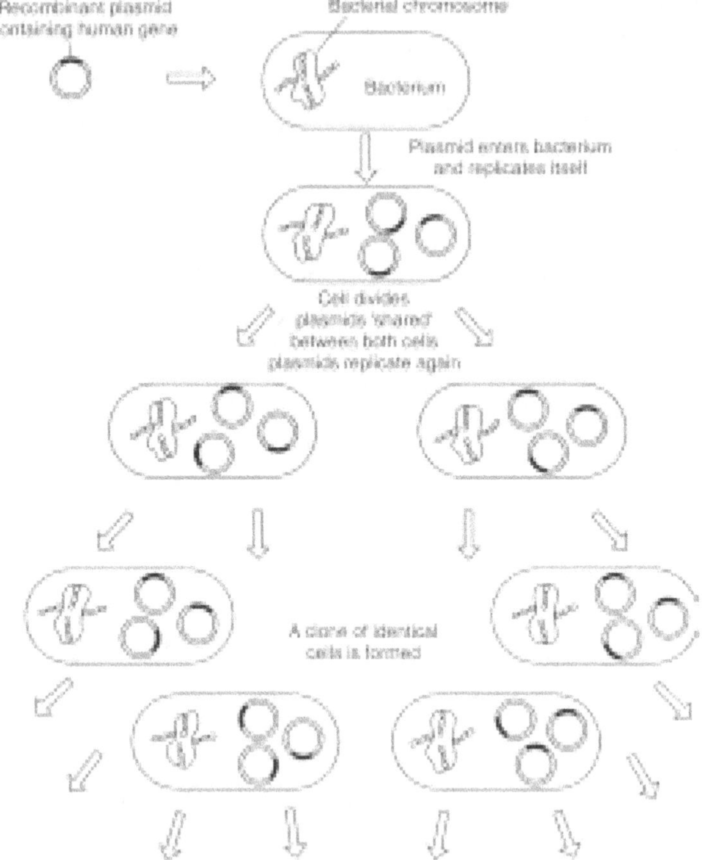

Figure 3.16   A cloning of recombinant plasmids containing human gene inside the bacterium

In a short span of time, a single bacterium will produce millions of descendants. A population of cells all derived from a single ancestor is known as a clone and all cells in a clone have the same genetic make-up. Thus a single bacterium carrying a recombinant vector will produce millions of identical cells, all of which contain the original human gene.

3.7.4 Selection of Right Clone

The next task for genetic engineers is to make the right choice for the clones of the bacteria which contain a recombinant plasmid with its human gene. There can be 3 types of bacteria confronted by genetic engineers:

- Bacteria infected by plasmids carrying a human gene (these are bacteria in which the scientists are interested).

- Bacteria having normal plasmids (i.e. they do not have human gene).

- Bacteria that have resisted the invasion by any type of plasmid (i.e. they are not transformed).

The next step, therefore, is to identify the clone having recombinant vectors. Suitable selection strategies have been applied to achieve this objective, and this is the most important step in DNA cloning. A simple but invaluable technique that helps in the checking process is replica plating (figure 3.17). The mixed population of bacteria is smeared across a plate of nutrients to individual cells. Each cell is allowed to multiply until it has produced a visible clone of cells. Thus there will be many distinct clones on the plate. Each clone is then separated into two portions, one is retained to grow in its ideal nutrients (master plate) while various experiments are carried out on the other portion (replica plate). Some of these experiments will kill some bacteria, but because their identical twins are kept alive and well in another place, every clone has copies of itself available for future investigations.

The bacteria which have no plasmids and those which have the normal plasmids need to be eliminated. Some types of plasmids carry genes that make bacteria resistant to certain antibiotics. One such plasmid is pBR322 which has the genes tet^r and amp^r for tetracycline and ampicillin resistance respectively. Any bacterium that contains

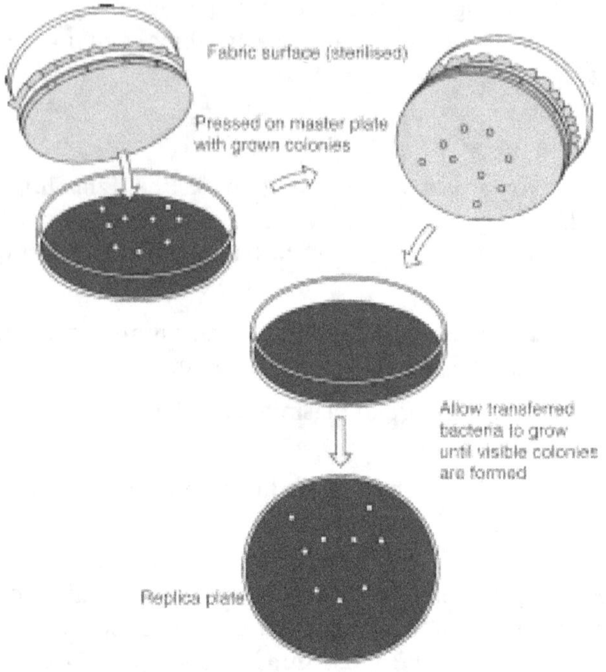

Figure 3.17   Preparation of replica plate with identical clones appear
in exactly the same position

the normal form of pBR322 will be unharmed by these antibiotics. However, bacteria with recombinant plasmids will not be able to protect themselves against tetracycline since the introduction of a human gene in the middle of tetracycline gene has destroyed its protective action. This forms the basis of a test (figure 3.18) to distinguish between bacteria with no plasmids (these will be killed by either antibiotic present in culture medium), those that have the normal plasmid (these can survive in the culture medium containing both antibiotics) and bacteria that have recombinant plasmids (these resist ampicillin but succumb to tetracycline).

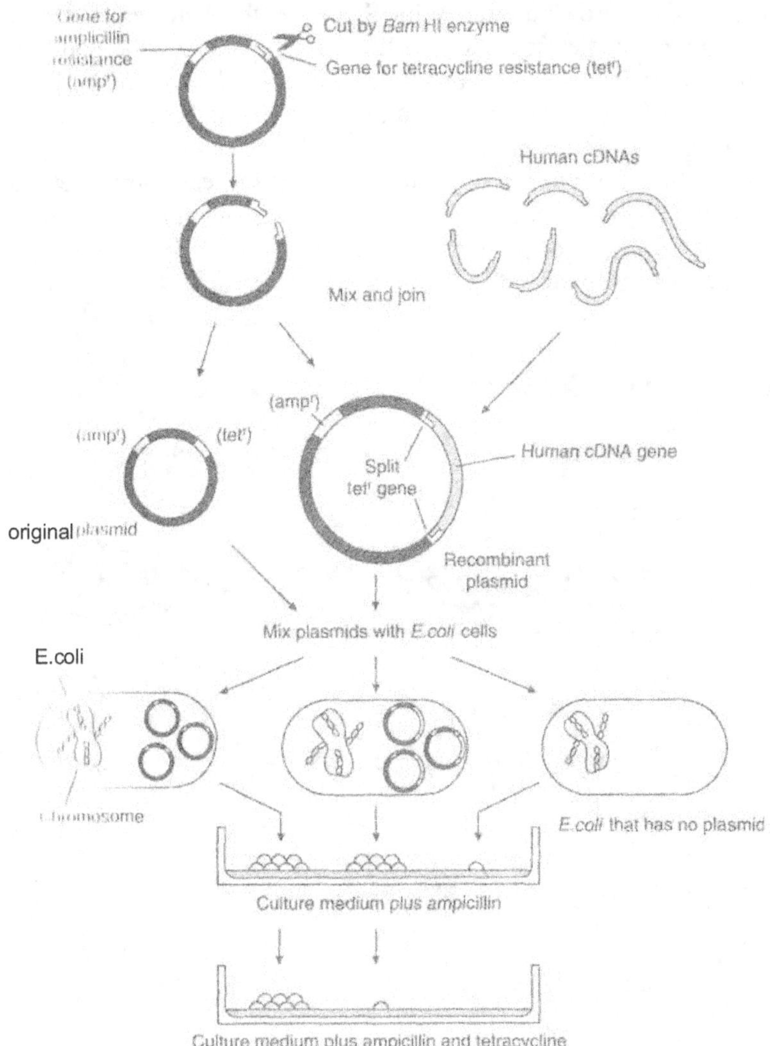

Figure 3.18   Identifying the bacteria which harbour recombinant plasmids
             containing human gene (cDNA molecule)

3.7.5 Colony Screening with Antibody for Pinpointing the Right Clone

After identifying the bacteria which harbour a recombinant plasmid/
human gene DNA molecule, the next step is to search for the clones

that manufacture the protein insulin. Since the whole procedure started with a mixed or heterogeneous collection of mRNAs (that code for many different proteins), only some clones will contain the gene responsible for production of insulin, while others contain different human genes. Radioactive antibody testing is one of the efficient and rapid methods for colony screening. The success of this method is due to two main facts—that antibodies can recognise and attach onto specific types of proteins (antigens), and that tiny amounts of radiation can be detected.

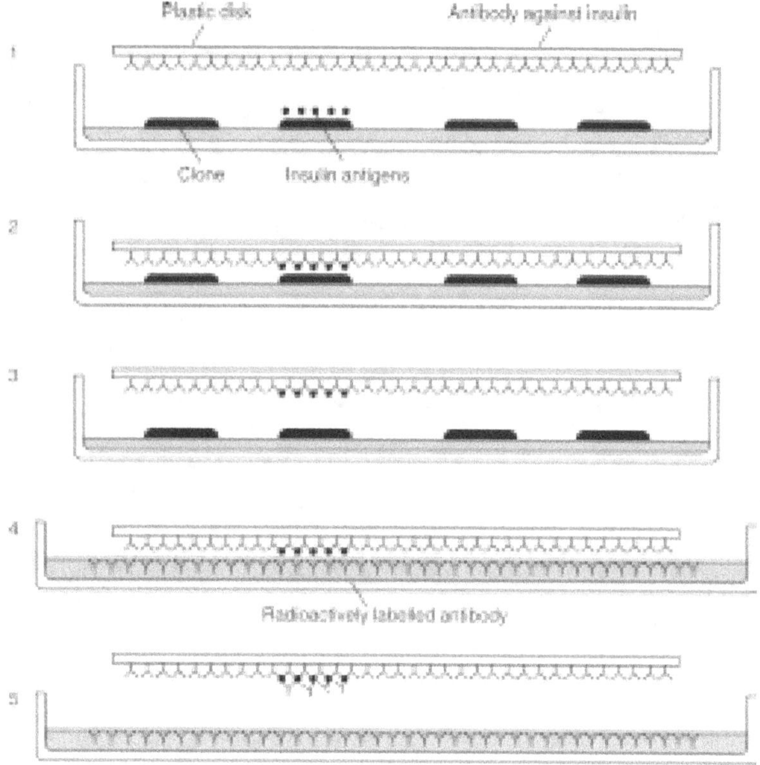

Figure 3.19   Pinpointing the right bacterial clone containing insulin DNA.
A plastic disk coated with antibody to insulin is first exposed to the contents of cells from each clone.
(1) Any insulin present in the cells is bound to the antibody.
(2) And thereby fixed to the plastic disk
(3) Radioactive labelled antibody to insulin is then applied to the disk in order to detect the presence of the protein (4,5).

Genetic engineers rely on the antibodies to search for bacterial clones that are producing insulin; by recognising insulin molecules, the antibodies stick to them like labels. These labels, like most molecules are too small to be seen directly. However, 'tagging' these labels with radioactive atoms helps them to become visible. These radioactively labelled antibodies emit very small amounts of radiation which produce a dark patch on photographic film. Figure 3.19 shows the combination of antibodies and radioactive labels that pinpoint to colonies of bacteria that manufacture insulin. Once insulin-producing bacterial clones are identified, they can be subcultured into an indefinite number of clones, producing a potentially unlimited supply of chemically pure, safe and effective insulin.

3.8 HUMAN GENOME PROJECT

The foremost application of genetic engineering has been in understanding the location, structure and function of each gene belonging to a wide variety of organisms ranging from bacteria to humans. For better understanding and manipulation of genes, an international approach is developed for the construction of the genetic and physical maps of the human genome.

Acquiring complete knowledge of the organisation, structure and function of human genome—the master blueprint of each of us—is the broad aim of human genome project which is undertaken by several research groups belonging to different countries. The imminent announcement of the completion of the human genome project (HGP), fifty years after the historic breakthrough by Francis Crick and James Watson marks both an end and a new beginning. It is the triumphant culmination of an extraordinary scientific collaboration that has involved 20 research groups from six countries—USA, UK, Japan, France, Germany and China. The US department of energy and the US National Institute of Health, Centre d'Etudes Polymorphism Humaine (CEPH) and Genethon (a factory of human genetics) in France and Welcome Trust Sanger Institute, England are the major contributory research institutions where this task was achieved. And it is the dawn of a new era in biological research, in which our growing understanding of the precise function and product of each of our 30,000–40,000 genes will bring important new advances not only in medicine but also in many other aspects of life on earth.

The idea that it would be possible to read the entire human genome was first raised in the late 1980s. The project was finally launched in 1990, with James Watson at its keen to harness what he called 'the great intellectual resources nurtured by the Medical Research Council and the Laboratory of Molecular Biology (UK)'. Watson sought the collaboration of the great British biologist Sir John Sulston and his team who were at that time working on *Caenorhabdits elegans*, the nematode worm.

One third of the three billion bases in the human genome have been sequenced (by the Welcome Trust Sanger Institute, UK) including chromosome 22, the first human chromosome sequence to be published.

The DNA used by HGP came from 12 anonymous volunteers. As humans are only 99.9 percent identical, the final sequence is described as a 'reference sequence' rather than a 'perfect sequence'. It is in fact, the 0.1 percent that makes us individual rather than identical clones.

DNA sequencing is the process of determining the exact order of the bases (A, T, C and G) in a piece of DNA. As it is not possible to sequence whole chromosomes (50–280 million bases) at a time, the genome is broken down into smaller pieces of DNA. Each piece is sequenced and then computers fit all the sequences together.

Once a chromosome has been chopped into convenient-sized chunks, these fragments are inserted into bacterial artificial chromosomes (BACs) and cloned in bacteria. The fragments are then mapped, so it is known which region of the chromosomes they came from. Each BAC is 'shot gunned' broken randomly into many small pieces. These fragments are cloned and then sequenced—automatic sequencing machines produce about 500 bases of sequence information from each fragment. The sequences are fed into a computer, which looks for overlaps at each end, to find neighbouring fragments, thus the whole BAC sequence can be assembled.

At most genomic laboratories, robotics and automation play a crucial role in large-scale sequencing and mapping. Robots such as the Beckman Biomerks 1000 and 2000 and Hewlett Packard ORCA system are adopted with increasing frequency in Human Genome Project at Lawrence Berkeley laboratory. A system proposed at General Atomics in La Jolla, California integrated a custom robot with several hardware and software architectures for speedy sequencing in a short span of time; this system was then adopted by most of the genome research laboratories.

The working draft of the human genome announced in 2000, covered about 90 percent of the human genome with an error rate of about 1 in 1000. Each fragment was sequenced four or five times. For the 'gold standard' version completed in 2003, each fragment has been sequenced about ten times, producing a final sequence with almost no gaps and error rate of less than 1 in 10,000 bases (99.99 percent accuracy). As one of the largest sequencing factories, the Welcome Sanger Institute at Hinxton, Cambridge, achieved an error rate of fewer than one mistake in 1,000,000 bases. In 2002, Rice became the first food crop to have its genome decoded. In the same year, the first public draft of the genome of mouse, an important model animal, was announced jointly by UK and US institutions.

"Fulfilling the true promise of the Human Genome Project will be the work of thousands of scientists around the world, in both academia and industry. It is for this reason that out highest priority has been to ensure that genome data are available rapidly, freely and without restriction." This is the opinion of the authors of the Human Genome Project sequencing paper. While Sir John Sulston, former Director of Welcome Trust Sanger Institute, UK, believes that we have now reached the point in human history where for the first time we are going to hold in our hands the set of instructions to make a human being.

Professor Allan Bradley took as Director of Sanger Institute in 2000 and recently announced a new five-year research programme that will involve translating the enormous amount of information included in our DNA into an understanding of how genes build and control the human body, and the roles that the products of human and pathogen genomes play in health and disease. The human genome sequence is utterly essential to the cancer genomic project. Since all cancers arise as a result of mutations in DNA, scientists are using the human sequence as a reference to find genes that are mutated in human cancers.

Significant output of the Human Genome Project is in the form of valuable information such as:

- The human genome contains 7164.7 million chemical nucleotide bases (A, C, T and G).

- Only about 5 % of the genome consists of gene sequences.

- The total number of genes is estimated at 30,000–40,000.

- The average gene consists of 3000 bases, but sizes vary with the largest known genes containing 2.4 million bases.

- Almost 99.9 percent of nucleotide bases are the same in all people.

- Chromosome 1 has the most genes (2968) and the Y (male)chromosome has the fewest (34) .

3.8.1 The Next Step

The complete human genome sequence is mankind's first step towards understanding how genes work, their complex interaction with each other and with the environment and their role in health and disease. In the coming decades, scientists will use data from the human and other genome projects to develop new technologies that will lead to healthier people and animals and less polluted environments. Genomics, the new area of science which reads and interprets genetic information, will bring

- Better diagnostics
- New targets for medicine
- Gene therapies to treat disease
- New vaccines
- Human-identical therapeutic proteins from animal milk.
- Improved grafts and cell repair system
- Safer food
- Healthier farm animals and improved meat and dairy quality
- Crops that can adopt to hostile and changing environments
- New way to collect and store solar energy
- New and better industrial biotechnology processes

To carry the Human Genome Project work forward, researchers have developed a new generation of scientific techniques, technologies and disciplines such as:

Transcriptiomics   It deals with the large-scale analysis of mRNA to allow the process of active gene expression. It looks at which genes are active in a cell.

Proteomics The study of proteome, which is all the proteins produced as a result of gene expression.

Key Concepts

- Vectors are self-replicating DNA molecules into which foreign DNAs or genes of interest are integrated to form recombinant DNA which are then inserted into appropriate host for gene cloning or expression of the gene of interest.

- The small fragments of DNA or RNA used to identify the complementary sequences in nucleic acid samples are called probes.

- Sequencing of DNA can be done by two methods

- (i) Maxam and Gilbert method in which chemical reagents are used to destroy specific nucleotides and break the DNA molecule at specific sites in 4 separate reaction mixtures.

- (ii) Sanger and Coulson method where four reaction mixtures containing single-stranded DNA, radioactively labelled oligonucleotides as primer, an enzyme DNA polymerase, all four types of deoxyribonculeotides, and low concentration of dideoxyribonucleoside triphosphates (dd NTP) are incubated for growing complementary chains.

- Human Genome Project was undertaken by 20 research groups belonging to six countries—US, UK, Japan, France, Germany and China.

- At Sanger Institute, UK, one third of three billion bases in human genome has been sequenced.

- The valuable information obtained form the HGP include the following:
  - human genome contains 3164.7 million bases
  - only about 5% of genome consists of gene sequences
  - the estimated no of genes is 30,000–40,000
  - the average gene consists of 3000 bases
  - almost 99.9 percent of bases are same in all people and chromosome 1 has 2965 genes while Y chromosome has 321 genes.

- Genomics is the new area of science, which reads and interprets the genetic information.

- Transcriptiomics, Proteomics, Metabolomics, Bioinformatics and Bio-nanotechnology are some new disciplines developed to carry forward the work of HGP.

Metabolomics   The technology that explores the metabolites (small molecular weight compounds) in a cell.

Bioinformatics   The application of computer and statistical techniques to the management of biological information such as DNA sequences, and the shapes of the biological molecules such as proteins. Also a vital tool used in pharmacogenetics.

Bio-nano technology   It involves a precise molecule-by-molecule control of products and by-products at the nanometre level (one thousand millionth of a metre).

Review Questions

1. Define genetic engineering or gene cloning. Explain the technique involved. Give its applications.

2. What are restriction enzymes? How do they work? Enlist any 10 restriction enzymes with their source and recognition sites.

3. Define cDNA library. Explain the steps involved in construction of cDNA library.

4. What is genomic library or gene bank? Explain the role of shotgun approach in developing gene bank.

5. Explain the process of enzymatic synthesis of DNA. Enlist the applications of synthetic oligonucleotides.

6. What is polymerase chain reaction? How are the genes amplified with help of PCR?

7. Define thermal cycler? What is the role of Taq DNA polymerase? Enlist the applications of polymerase chain reaction.

8. Define vector. What are the types of vectors? What are the properties of good vectors?

9. Describe different types of plasmids and phage vectors used for cloning.

10. Define probes. What are the types of probes? How they are useful?

11. Define DNA sequencing. Explain the methods of DNA sequencing.

12. Explain how DNA sequencing is useful in genetic engineering.

13. Define rDNA technology. Explain the processes involved in commercial production of insulin.

14. Define Human Genome Project. What are the objectives of HGP? Enlist the research groups and their contributions in HGP.

15. What is Human Genome Project? What are the significant achievements of HGP? How is it useful for welfare of human beings?

16. Write short notes on:

   a. Gene

   b. Clone

   c. Recognition site

   d. Bam HI

   e. Palindrome

   f. Blunt and sticky ends

   g. Reverse transcriptase

   h. Introns

   i. Contribution of H.G. Khorana

   j. Taq DNA polymerase

   k. Cosmid

   l. DNA Virus

   m. RFLP (Restriction fragment length polymorphism)

   n. Dideoxynucleotide

   o. Sequence of nematode worm

   p. FRIGIDA, the gene from *Arabidopsis*

   q. Replica plating

   r. Role of radioactive antibody in colony screening

   s. Nonbacterial proteins

   t. Genomics and bioinformatics

4

TRANSFECTION OR GENE TRANSFER TECHNOLOGY

4.1 INTRODUCTION

Dramatic progress in the field of genetic engineering has made it possible to produce sufficient amount of food, fibre and fuel for the ever-increasing population of the developing country. The term **transfection** specifies the introduction of the desired DNA segment or gene into an organism (bacterium, yeast, plant or animal) with such skill and efficiency that a transgenic organism is produced showing improved characteristic features which are beneficial to human welfare.

Introduction of foreign DNA into suitable plants and animals and their stable expression generates transgenic plants and transgenic animals. The transferred gene is called the **transgene** and the whole process is called transgenesis. Since it involves transfer of DNA or gene, the process is also known as **gene transfer technology**.

4.2 TRANSFECTION METHODS

Once the desired genes are identified and selected, they are transferred to plants or animals using various possible transfer methods. The technique differs in different organisms. Plant cell has three physical barriers in the form of cell wall, plasma membrane and nuclear membrane while animal cell lacks cell wall so it has plasma membrane and nuclear envelope as physical barriers. A variety of gene transfer techniques have been used to introduce transgenes in plants or animal cells. Depending upon the objective of transfection in the target cell, any one of the following methods or their combination can be used to transfer the gene into the cell. A number of approaches are currently being explored for gene transfer and their success rate is variable.

1. Agrobacterium-mediated DNA transfer This method is used to transfer the DNA in plant cells to express transgene to produce transgenic tissues and finally regenerate transgenic plants. *Agrobacterium tumefaciens* is a soil-borne bacterium. It has the capacity to produce crown gall tumours in dicotyledonous plants (figure 4.1). This tumour-inducing capacity of *Agrobacterium* is due to its large extra-chromosomal plasmid known as the tumour-inducing plasmid (Ti plasmid). Ti plasmid has the DNA sequence called T-DNA (transferred DNA), which is responsible for transformation of plant cell during the infection caused by *A. tumefaciens*.

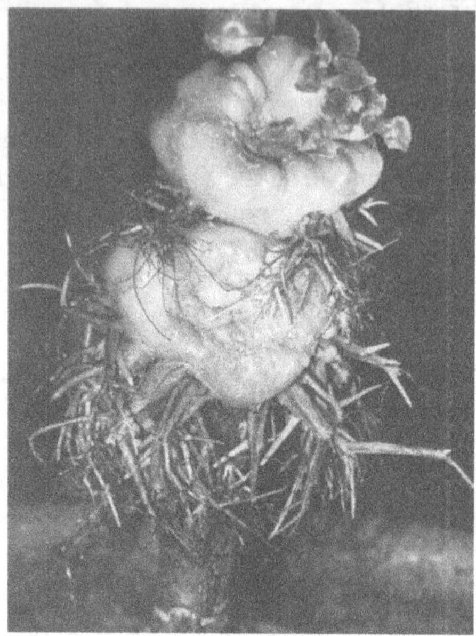

Figure 4.1   Crown galls, cancer-like growths on plants caused by *Agrobacterium tumefaciens.* The Ti (tumour–inducing) plasmid of this bacterium inserts itself into plant cells and causes them to grow and divide in an uncontrolled way.

In addition to T-DNA, Ti plasmid has the following important functional regions (figure 4.2).

   i. vir gene that regulates the transfer of T-DNA into plant cells.

   ii. oc region i.e. Opine catabolism region which produces enzymes necessary for the utilisation of opines (These are plant metabolites synthesised under the control of T-DNA, which can be specifically catabolised by the inciting bacteria and thus used as carbon and nitrogen source).

   iii. tra (conjugative transfer) region functions in conjugative transfer of plasmid.

   iv. ori region which is responsible for origin of replication for multiplication in bacterium.

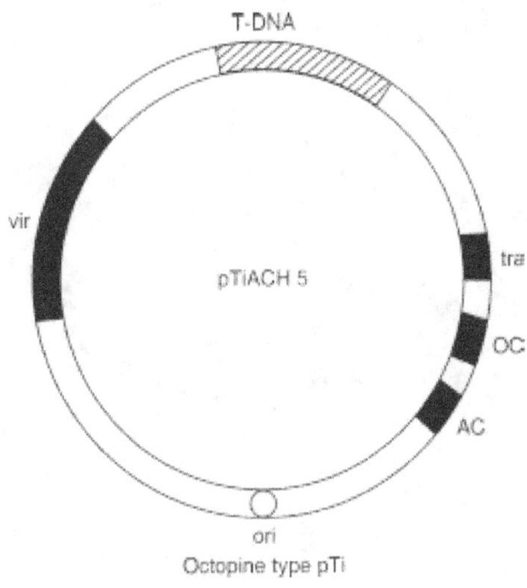

Figure 4.2   Structural features of Ti plasmid

Agroinfection   In this approach the plant cell is infected with the recombinant vector which is to be prepared by introducing the viral genome into the T-DNA of Ti plasmid and using the *Agrobacterium* (figure 4.3). When the recombinant plasmid is co-cultured with plant cells the process is called agroinfection. For the successful transmission of the infection, geminiviruses which have to be transmitted by an insect vector (leaf hopper) are usually employed.

Gemini viruses have single-stranded circular DNA infecting a variety of monocot and dicot plants. They include maize streak virus (MSV) causing yellow streaks in maize leaves and wheat dwarf virus (WDV) infecting the seedlings of wheat.

Gene transfer mediated through *Agrobacterium* has been demonstrated in tissue explants of species like *Asparagus, Allium cepa, Dioscorea bubbifera, Chlorophytum,* etc. Similarly imbibition of the seeds of *Arabidopsis thaliana* (thale cress, a relative of the mustard and cabbage families) in a fresh culture of *Agrobacterium* leads to stable integration of T-DNA in the *Arabidopsis* genome.

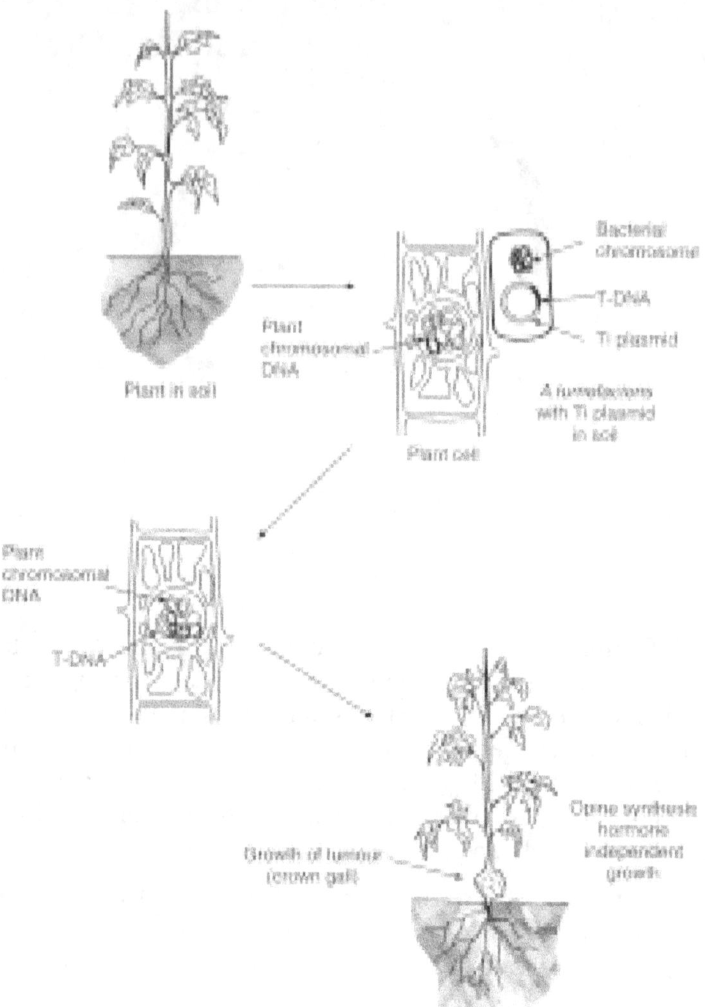

Figure 4.3   Induction of plant tumour by *Agrobacterium tumefaciens* carrying Ti plasmid

2. Gene transfer through RNA viruses (Retroviral infection) This method is applicable for transfection of plant and animal cells. Plant virus genomes do not integrate into plant genome. As a result, they cannot be used to produce stable and heritable transformations. But they can be used to express transgene to either improve the phenotypic performance of host plants or to produce large quantities of valuable proteins. Tobacco mosaic virus (TMV) used to infect tobacco and Brome mosaic virus (BMV) used to infect several species of Graminae including barley, have RNA as genetic material. Both RNA viruses are used to transform the plant cells for expression of transgene.

Animal viruses are also known as retroviruses (e.g. HIV, HAV, influenza virus, poliomyelitis virus and MOMULV (Moloney Murine Leukemia Virus). Transformation of mouse embryos (4–8 cell stage) can be brought about by infecting the cells of pre-implantation mouse embryo with MOMULV. This may be done either by co-cultivating 4–8 cell pre-embryos with virus producer cell or by injecting virus into blastocyst *in vitro*.

3. Pollen transformation This method can be applied for DNA transfer in plant cells. This can be achieved by soaking the pollen grains in DNA solutions prior to their use for pollination. Though this method is simple and has high applicability, it lacks practical evidences.

4. DNA transfer via growing pollen tube In this method, the stigma of a flower is cut off sometimes after pollination, and DNA solution is applied onto the cut surface. Using this method transformation of rice and barley can be done. This technique is simple, easy and gives prominent results.

5. DNA uptake by mature zygotic embryos This method is employed for transfection of plant cells. The zygotic embryo is present in the seed. Grinding the seeds in chemicals like CCl_4 or cyclohexine can isolate the dry embryos of wheat, barley, ray, pea or bean. The imbibitions of the dry embryos in a solution containing the desired gene can take up DNA and can show the expression of the desired gene. The imbibed embryos can be germinated on appropriate selective media to isolate the transformed embryos.

6. Protoplast fusion This is a versatile technique in which different genomes of different genera and species can combine to form somatic hybrids and cytoplasmic hybrids (cybrids). The technique has the potential to overcome the sexual incompatibility barrier between plants. Details of the method are discussed in Section 2.13.

7. Calcium phosphate-mediated gene transfer The technique is applicable to transfect animal cells, which involves the admixture of isolated DNA with solution of calcium chloride and potassium phosphate under controlled conditions which allow fine precipitate of calcium phosphate to be formed. Cells are then incubated with precipitated DNA either in solution or in a tissue culture dish. A fraction of cells will take up the calcium phosphate–DNA precipitate by endocytosis. This is a general method for introducing any DNA into mammalian cells.

8. Use of polyethylene glycol for DNA transfer This process of DNA transfer using polyethylene glycol (PEG) is applicable to protoplasts only. PEG is one of the chemical fusogens used to induce the fusion experiments. PEG causes the isolated protoplasts to adhere to one another and leads to tight agglutination followed by fusion of protoplasts. After exposure of the protoplasts to exogenous DNA in presence of PEG and other chemicals, PEG is removed. Intact viable protoplasts are then cultured to regenerate cell walls and colonies of transformed plant cells.

9. DEAE-dextran-mediated gene transfer DEAE-dextran can be used in the transfection medium in which DNA is also present. DEAE-dextran is soluble and it does not precipitate. It is a polycationic high molecular weight compound, which probably acts by mediating in some unknown way the productive interaction between negatively charged DNA and the components of cell surface in endocytosis. This technique has been previously used to increase the efficiency of viral infection of cell-lines.

10. Use of liposomes for DNA transfer Introduction of DNA into cells via liposomes is known as lipofection. Liposomes are the artificial phospholipid vesicles; they can be preloaded with DNA and then fused with protoplast to deliver their contents into cytoplasm. Using this technique, it became easy to transfer DNA into animal cell, protoplast of yeasts, plants and bacteria (figure 4.4).

11. Biolistic or particle gun In this method, 1–2 μm tungsten or gold particles, are coated with the DNA to be used for the transfection of plant cells or nuclei. To accelerate the velocity of particles, the pressurised helium gas or the electrostatic energy released by a droplet of water exposed to voltage can be used. Since this pressurised helium gas is released from a device which uses a blank cartridge in a modified firing mechanism to provide the energy for particle acceleration, this technique is given the name 'particle gun'. This method is also called biolistic or ballistic method of DNA transfer and it can be used for any plant cell, leaves, root section, embryos, seeds, pollens, etc.

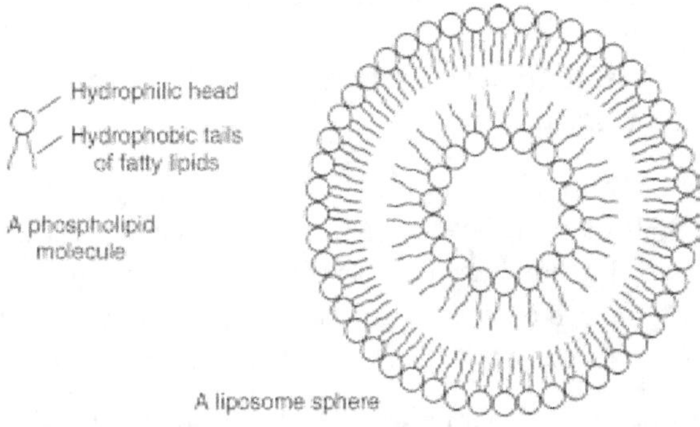

Figure 4.4   Molecules of phospholipid aggregated to form liposome—a spherical structure used to transfer genetic material inside the cell

12. Microinjection   Though this technique is applicable to plant and animal cells to transfer DNA, it is more commonly used to transfect animal cells. In the case of microinjection, the DNA solution is injected directly inside the cell using capillary glass micropipettes with the help of micromanipulators of a microinjection assembly (figure 4.5).

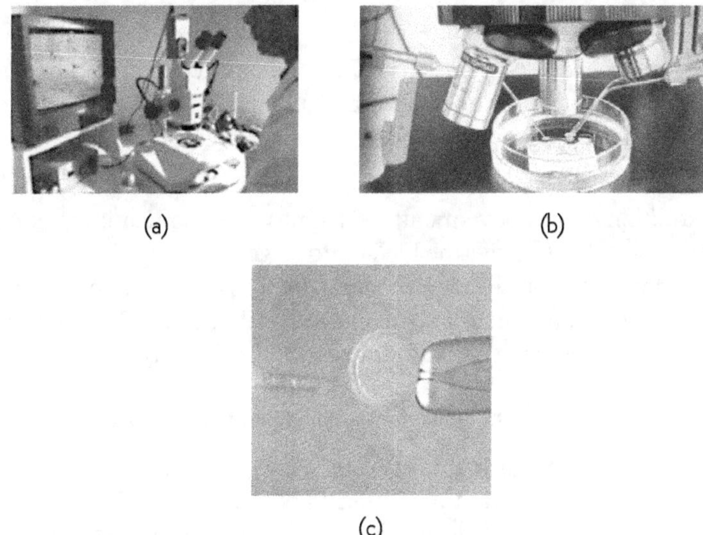

(a)   (b)

(c)

Figure 4.5 (a,b and c)   Assembly for embryo manipulation by microinjection

This method is ideally useful for producing transgenic animals quickly. Using this technique, DNA is introduced into fertilised egg before the fusion of male and female pronuclei. This method has been successfully utilised to generate transgenic mice, chicken, cow, pig, sheep and rabbit. The refined technique of microinjection is successfully used in the introduction of DNA into the protoplasts of tobacco, alfalfa, *Brassica* sp., etc.

13. Macroinjection This technique involves the injection of plasmid DNA into the lumen of developing inflorescence using a hypodermic syringe. Since the needles used for injecting DNA have a greater diameter than the cell diameter, the technique is called macroinjection. This method is successfully used to transform the rye plant.

14. Electroporation This method can be used to introduce desired DNA to a variety of cells like animal cell lines, plant protoplast, yeast and bacterial protoplast. The technique involves introduction of DNA into cells by exposing them for very brief periods to high voltage electric pulses, which is thought to induce transient pores in the plasmalemma.

15. Fibre-mediated DNA delivery In this approach DNA is delivered into the cell cytoplasm and nucleus by silicon carbide fibres. This method was successful with both maize and tobacco suspension culture cells.

16. Laser-induced DNA transfer This technique involves the use of laser for transfection of animal and plant cells. Laser punctures transient holes in plasma membrane through which DNA may enter into the cell cytoplasm.

In addition to the above-mentioned approaches, sonication method can also be used to transfer the DNA into target cells. In this technique the sound waves are utilised to produce the transient pores on plasma membrane through which DNA can enter into the cell. This method may become popular in future because of its simplicity.

4.2.1 Transfection Techniques for Animal Cells or Embryos

Introduction of DNA into animal cells/embryos is the critical step in the production of transgenic animals. There are several techniques that can be applied to achieve this goal. Some of them include:

   i. Calcium phosphate precipitation

  ii. Microinjection

 iii. Retrovirus infection

 iv. Lipofection

  v. Particle gun delivery and

 vi. Electroporation

These techniques have already been described in the previous sections. In addition to these transfection methods, transgenic mice are produced by embryonic stem (ES) cell transfer technique which is described in detail.

Embryonic stem (ES) cell transfer technique Embryonic stem (ES) cells are found in the mammalian blastocyst, which is an early stage of embryonic development comparable to blastula of other animals. ES cells have virtually unlimited powers of differentiation. The mammalian blastocyst is composed of two distinct parts, a thin outer layer of cells and an inner mass of cells. The outer layer makes up the trophoblast, which gives rise to most of the extra-embryonic membranes characteristic of mammalian embryos. The inner surface of the trophoblast contacts the cluster of cells, called inner cell mass, that projects into the spacious cavity known as blastocoel. The inner cell mass is the part of blastocyst that gives rise to the cells that make up the mammalian embryo. The inner cell mass contains the embryonic stem (ES) cells, which are capable of differentiating into all of the various tissues that a mammal comprises. Thus ES cells have tremendous potential for differentiation hence are called pluripotent cells which can be isolated from blastocyst (pre-implantation embryos) and cultured *in vitro* under controlled conditions that accelerate the growth and proliferation of cells for a long time to permit the various manipulations for gene transfer.

The cultured pluripotent ES cells are transfected with the appropriate transgene by a suitable transfection method. Transfected ES cells are identified and selected, usually by applying a selectable marker gene, and cloned.

ES cell transfer technique is significantly employed for specific transgenic animal strains that are created for specific experimental or biomedical needs. One such transgenic animal model is knock out

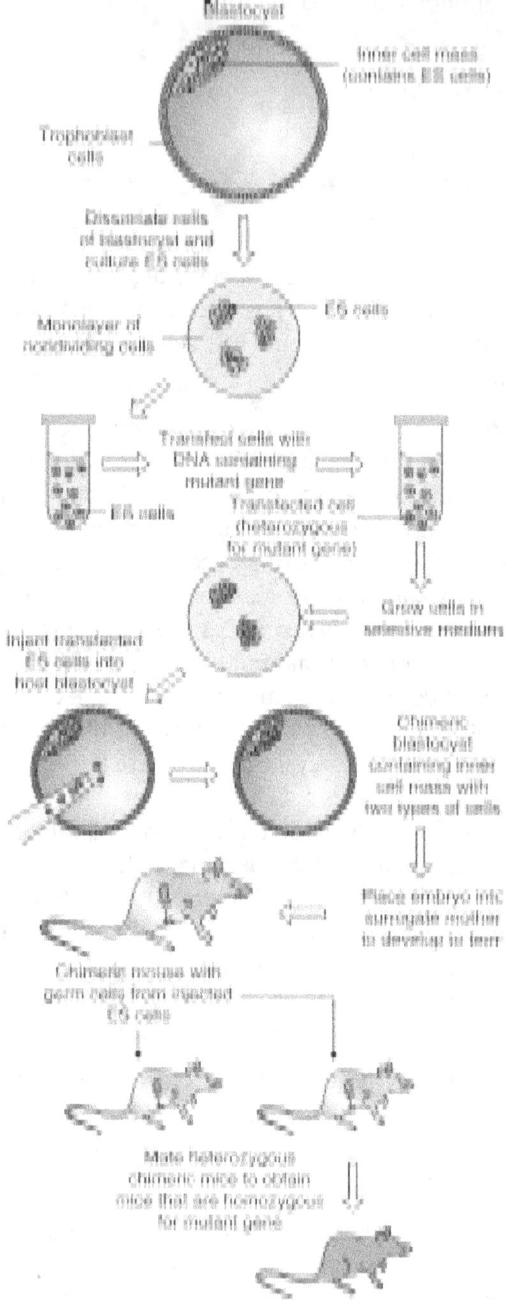

Figure 4.6 Steps involved in formation of knockout mice

mice, these animals can provide a unique insight into genetic basis of a human disease as well as a mechanism for studying the various cellular activities in which the product of a particular gene might be engaged. For creating knockout mice (figure 4.6), the cultured pluripotent ES cells are transfected with DNA fragments containing a mutant allele of the gene to be knocked out, as well as genes that can be used to select cells that have incorporated the altered DNA into their genome. Of those cells that take up the DNA, approximately one in ten thousand undergoes a process of homologous recombination in which the transfecting DNA replaces the homologous DNA sequence that contains the normal alleles. Using this procedure, ES cells that are heterozygous for the gene in question are produced and then selected on the basis of their drug resistance.

The next step is to inject a number of these ES cells into the blastocoel of a mouse blastocyst and implant the injected embryo into the oviduct of a female mouse that has been hormonally prepared to carry the embryo to complete the development. As the embryo develops in its surrogate mother, the injected ES cells join the embryo's own inner cell mass and contribute to the formation of embryonic tissues, including germ cells of gonads. The offspring of such mice are heterozygous for the gene in question and can be mated with one another to produce offsprings that are homozygous for mutant allele. These mice now lack a functional copy of the gene. The knock out mice produced by this technique, which lack a functional p53 gene (a tumor-suppressor gene encodes the protein having a molecular weight of 53 kiloDalton) are almost certain to develop malignant tumours.

4.3 TRANSGENIC ANIMALS

A transgenic animal is a fertile animal, which carries a gene or genes introduced by one or other technique of transfection. Farm animals like cattle, sheep, goat, poultry, pigs and fish are utilised as the major source of food, milk, wool, leather and many other useful by-products. In the developed world, animal production is highly intensified and technologically driven. Similarly the biotechnologists of the developing world are achieving the goal of artificial breeding and control of animal reproduction using artificial insemination, super-ovulation, embryo transplantation and gene transfer techniques to increase the animal products to meet the demands of the ever-increasing population.

Selective breeding is the method used by animal breeders to improve the genetic features of livestock animals. This process is time-consuming especially in larger animals with long gestation period, and can take many years to establish desired phenotypic changes. The gene transfer technique made it possible to introduce cloned genes into fertilised eggs and their successful implantation into receptive females to obtain the progeny carrying cloned gene has opened up new avenues for the development of transgenic animals.

The first transgenic animals were produced in 1981 by Ralph Brinster of the University of Pennsylvania and Richard Palmiter of the University of Washington. They succeeded in introducing a gene for rat growth hormone(GH) into the fertilised eggs of mice (figure 4.7a). The subsequent progeny were all much larger than the parents. This larger 'super-mouse' weighed 44g while the smaller uninjected controlled mouse had a weight of 29g. The transgenic mouse is useful for various experimental studies, which has acquired novel genetic material by artificial means rather than by the normal route of sexual reproduction (figure 4.7b).

Subsequently the intention of biotechnologists is diverted to produce transgenic farm animals from which the mankind can derive greater benefits. Many of the farm animals are improved for their meat production ability while some of them are improved for milk yields and quality, and disease-free status.

The main objectives for improved animal breeding programmes coupled with this new technology of gene transfer are given below.

- Efficiency of meat production

- Improved quality of meat

- Milk quality and quantity

- Egg production

- Wool quality and quantity

- Disease resistance in animals

- Production of low-cost pharmaceuticals and biologicals

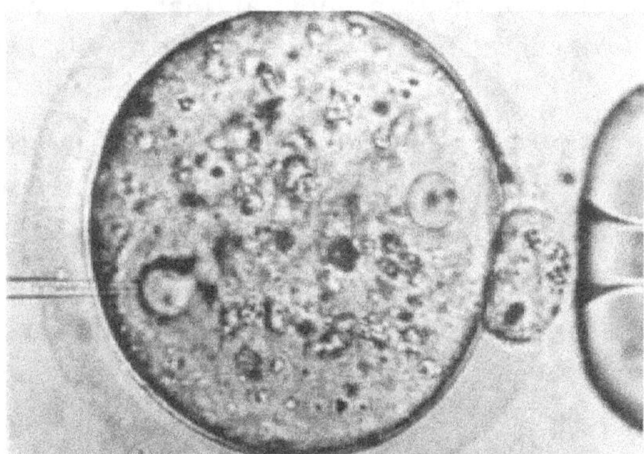

Figure 4.7a Microinjection of DNA into the nucleus of a recently fertilised mouse egg

Figure 4.7b Transgenic mice. The larger mouse developed from an egg injected with rat growth hormone gene

4.3.1 Animals Produced by Gene Transfer Technology

Transgenics have been produced in a variety of animal species, e.g. mice, rabbits, pigs, sheep, goat, cattle, poultry, fish, amphibians, insects and nematodes by any of the techniques of transfection like microinjection, retroviral infection, calcium phosphate-mediated transfection. The exploitation of transgenic animals on a commercial basis can be classified into two groups:

- The production of transgenic animals for increase in the quantity of conventional products like meat, milk, leather, wool, etc.

- Creating transgenic animals for synthesis of nonconventional products i.e. animals acting as bioreactors to synthesise valuable proteins of pharmacological importance like interferons, antitrypsin, etc.

I. Transgenic Mice   Mice are the appropriate animals selected for studies on gene transfer due to several of their favourable characters like short generation time, production of several offspring per pregnancy, easy *in vitro* fertilisation, successful embryo culture, production and maintenance of ES cell lines, etc. As a result, transgenic mice have been used as a model for transgenic experiments involving the introduction of oncogenes into embryos and adults to induce tissue specific tumours helping in the study of human cancers. Cultured cells that are immortalised by oncogenes can produce pharmaceutically significant proteins. Thus transgenic mice play a significant role in studying fundamental problems of development and disease in mammals. The production of genetically modified mice triggers the process of gene transfer technology (figure 4.8) with modified approach to develop other transgenic animals.

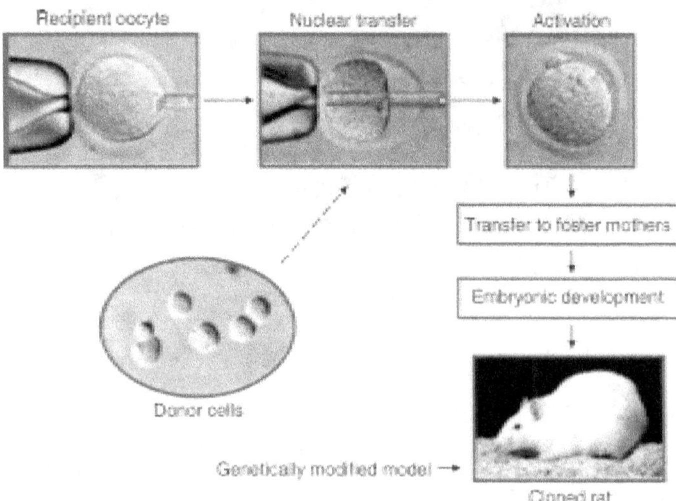

Figure 4.8   Gene transfer technology to produce genetically modified rat

2. Transgenic Sheep   Gene transfer technology is applied to sheep to produce trangenic sheep which are able to achieve better growth and meat production as well as to serve as bioreactors. Human growth hormone gene is introduced in sheep for promoting the growth and meat production. The transgenic sheep showed amelioration in body weight, redistribution of subcutaneous fat and feed efficiency. However, they also showed side effects like skeletal and joint defects, ulceration and infertility which may be due to long-term administration of elevated levels of growth hormones.

Bacterial genes, cys E and cys M, are concerned with biosynthesis of cysteine amino acids involved in formation of keratin protein found in wool. Both these genes are identified, cloned and introduced in sheep to increase wool production and to improve the quality of wool.

The role of transgenic sheep as a bioreactor is a fabulous achievement. Human genes responsible for synthesis of blood clotting factor IX and alpha-1-antitrypsin have been introduced in sheep and expressed in mammary tissues. Such approach is called **molecular or gene farming**.

In 1990 Tracy, the transgenic ewe was born in Scotland, and could produce a human protein in her milk for human therapeutics. It was hoped that the protein alpha-1-antitrypsin would prove useful in combating cystic fibrosis.

3. Transgenic pigs   The objective of gene transfer in pigs is to increase growth and meat production and to act as bioreactors. Human gene responsible for growth hormone synthesis and introduced in pigs showed improved growth and meat production similar to that of transgenic sheep, but they also showed certain undesirable characters. Human respiratory protein haemoglobin and specific immunoglobulins are easily recovered from either milk or serum of transgenic swine transfected with human haemoglobin and antibody synthesising genes. These protein products can be used in blood transfusion and disease diagnosis.

In January 2002, an Edinburgh-based therapeutics company announced the birth of a litter of transgenic pig clones. A gene that would normally lead pig organs to be rejected by human body has been switched off in the cloned pigs, which in theory means that their tissue is a closer match for transplant operations. Cloning transgenic pigs would allow a ready supply of organs to be produced. Pigs are regarded as the most suitable animals to be bred for heart transplant because a pig's heart is about the same size as a human heart, and pig heart valves have been used in human heart surgery for over a decade. Much research still remains to be done especially regarding the possibility of pig tissue infecting patients with pig viruses. But the pig clone is the first step towards providing animal organs and tissues for human transplants (xenotransplantation).

4. Transgenic fish   In fishes the eggs are fertilised externally, thus eliminating many of the complicated techniques required in mammals to harvest ova, fertilise them and then introduce the embryos into foster mothers. Therefore the rate of fish transgenesis can be as high as 70%. The commercially important fish like Atlantic salmon, catfish, goldfish, *Tilapia*, zebra-fish, common carp, rainbow trout, etc. are transfected with growth hormone, chicken crystalline protein, antifreeze peptides and glycoproteins, *E.coli* β-galctosidase and *E.coli* hygromycin resistance gene. Transfection is achieved either by microinjection of desired gene into ooplasm or by electroporation. Transgenic fish showed increased cold tolerance and improved growth and it is the quantity and quality of fish proteins as well as its preservation are the factors affecting the economic value of fish.

5. Transgenic cattle   Gene transfer technology is also applied to produce transgenic cattle to improve milk and meat production. The intention of genetic engineers is to improve the properties and proportion of casein, to increase the content of lactose and butterfat in milk, and to induce the resistance to viral and bacterial pathogens during the transgenesis in cattle. Several human genes have been successfully introduced in cows and expressed in the mammary tissue. Parallel results are obtained in transgenic goats, where they are used as bioreactors.

4.4 TRANSGENIC PLANTS

The human race is totally dependent on agriculture and as world populations continue to expand there must be continuous reassessment of agricultural practices to optimise their efficiency. Since early times human beings have sought to improve the quality and productivity of agriculturally important plants. This was done by selection and traditional breeding procedures that were painstakingly slow and difficult. Traditional breeding programmes involve sexual crosses which resulted in the high quality of present day food plants such as wheat, rice, corn, potato, etc. More recently, biotechnological approaches have been applied to these plants to create genetic variations that are beneficial for mankind. Culture of desired plant cell by *in vitro* technique and production of somatic hybrids and cybrids by protoplast fusion technology often resulted in the production of hybrid plants between sexually incompatible species. However, the most potent biotechnological strategy applied for the transfer of specifically constructed gene assemblies through various transfection techniques constitutes genetic engineering. The plants produced through genetic engineering contain gene or genes usually from an unrelated organism. Such genes are called transgenes, and the plants having transgenes are called transgenic plants.

Many of the methods developed with animal-cell gene cloning and gene transfer have already been shown to work in plant cells. Plant genes have been isolated and successfully incorporated into bacteria to create recombinant DNA molecule. Furthermore, some of these recombinant molecules have been transferred back into eukaryotic plant cells by any one of the transfection methods described earlier, where the cloned gene has been incorporated in host cell DNA and expressed resulting in a phenotypic change to suit various human needs.

Genetic engineering methods allow the transfer of a single gene into plant cells and this should extend later to multiple genes and indeed to whole biochemical pathways. The goals for the application of genetic engineering to plants are as follows:

1. Improved resistance to specific herbicides Application of herbicides to kill the plant weed species provides a growth advantage to commercial crop plants. Herbicides are designed to disrupt the growth of certain weed species without affecting the particular crop plant. However, there is increasing opposition to continued use of such chemical compounds on the grounds of environmental and human health considerations. Therefore it is necessary to develop less toxic, efficient and safer herbicides. Recently several such herbicides have been manufactured but only a very few of these fulfil the ideal requirements of a truly selective herbicide in controlling all plants except the cultivated crop. Examples of selective herbicides are atrazine in maize, dichloromethyl in wheat, auxin-type herbicide in many monocot plants and sulfonylurea and imidazoline herbicides in various cultures. Most of the modern herbicides interfere with amino acid biosynthesis in plants. Once the mode of action of a herbicide is established, the transfer of resistance to herbicide via modification of the target site is possible, in addition to detoxification of herbicide, for conferring herbicide resistance.

An ideal herbicide should have following qualities (i) It should control the growth of all plants species except the desired crop, (ii) It should have low mammalian toxicity, (iii) it must have higher degree of environment safety, (iv) it should rapidly degrade by soil microorganisms, and (v) they must have minimum persistence in soil. None of herbicides used is having all above-mentioned characteristics. Since weed control is an indispensable aspect of modern agriculture, biotechnology will make it possible to produce the herbicide having desired qualities in the near future.

2. Improved resistance to insect pest Insect pests cause damage through destruction and spoilage of standing crops, and stored agricultural products. Recent survey shows that about 30% of agricultural products are destroyed by pests every year. Control of insects in crop plants depends largely on synthetic chemistry. However, an alternative has been available for more than 30 years which is a biological insecticide from the bacterium, *Bacillus thuringiensis* (Bt). But the use of *B. thuringiensis* sprays is limited because of low stability of the protein in the field, relatively high production cost and the limited insecticidal spectrum. With the arrival of methods to engineer crop genetically, it is possible to make broader use of natural insecticides. Moving *Bt* insecticidal gene from the bacterium to plants allows crop protection from insect attack without the aid of externally applied chemicals.

Insect resistant plants contain either a gene from *B. thuringiensis* or the cowpea trypsin inhibitor gene. The choice of a *Bt* endotoxin is based on extensive knowledge about insecticidal crystal protein (cry protein). The gene called cry gene present in *B. thuringiensis* produces a protein that forms crystalline inclusions in bacterial spores. When ingested by a susceptible insect, a combination of high pH and the enzyme proteinase of the insect's midgut, processes them hydrolytically to release the core toxic fragments. The effect of these fragments is seen within minutes of ingestion, beginning with midgut paralysis and ending with disruption of midgut cells of insect. *Bt* toxin activity has been against many species of insects within the orders of Lepidoptera, Diptera, and Coleoptera.

Cry genes have been successfully transferred into tobacco, potato, cotton and tomato. Typically, truncated cry genes are used for the production of transgenic plants since the level of expression of the complete genes in transgenic plants is extremely low.

Many plants have evolved natural defence mechanisms against herbivorous insects. One such mechanism is the synthesis of proteinase inhibitors. In contrast to *Bt* endotoxin, these proteins have anti-metabolic activities against a wide range of insects. In 1987, Hilder *et al* in England had produced insect resistant tobacco plants through introduction of a serine proteinase inhibitor gene derived from cowpea. Plants expressing very high level of cowpea trypsin inhibitor showed decreased insect damage. The genes responsible for the product that inhibits the action of digestive enzymes like cowpea trypsin inhibitor, serine, cysteine protease inhibitor and proteinase inhibitor II, etc. have been identified, isolated and transferred to crop plants. Such transgenic plants show promising results against different insect pests belonging to order Lepidoptera, Coleoptera, etc. by improving their insect resistivity.

3. Improved resistance against microbial diseases   Microbial diseases caused by fungi and viruses are the major factors limiting crop productivity worldwide, with continuing huge losses set against large cash inputs for pesticide treatment. In improving the resistance of plants against fungal pathogens, an extensive research is being carried out on the genes that encode enzymes involved in the synthesis of substances toxic to fungi and for genes encoding proteins with direct inhibitory effect on the growth of fungi.

Under certain conditions, both microorganisms and plants produce low molecular weight antimicrobial substances. In plants such compounds are called phytoalexins which are synthesised locally and accumulate after exposure to pathogen. In *Vitis vinifera* and *Picea sitchnesis*, stilbenes are synthesised in response to fungal infection as a protective strategy. Tobacco contains the precursors for the formation of stilbenes but lack the enzyme stilbene synthetase. After transfer of a stilbenes synthatase gene from grapevine to tobacco transgenic plants expressed the foreign gene in response to fungal infection.

Cell wall of many fungi contain chitin and β-1,3-glucans as the major structural polysaccharides, these components are broken down by endochitinase and β-1,3-endoglucanase to inhibit the growth of fungi. The genes responsible for synthesis of these hydrolases (proteins) can be introduced into plants for fungal resistance.

Similar to the fungal infection, there are several plant viruses, causing diseases in crops. Engineering strategies are based on the introduction of virus-derived or a virus-targeted sequence into the genome of a potential host plant. These sequences interfere with the specific stages in the viral infection cycle. Coat protein (CP)-mediated resistance to virus has been one of the successful approaches to develop the transgenic plants with viral resistance. One such example is introduction of tobacco mosaic virus (TMV) coat protein gene into the genome of tobacco plant. Subsequently, CP genes isolated from different single-stranded RNA plant viruses have been cloned into plant expression vectors and transferred into various crop plants (tobacco, tomato and potato) using the *Agrobacterium tumefaciens-*mediated plant transformation system.

4. Modification in post-harvest characteristics Diseases and pests, bruising on soft fruits and vegetables, heat and cold storage, over-repeness, loss of flavours and odours, etc. lead to great deal of losses during storage and transport of crops. Most of these physiological changes are due to endogenous enzyme activity. Genetic engineering has made it possible to slow down these activities. In the tomato the enzyme polygalacturonase breaks down the cell wall constituent—pectin, leading to softening of fruit during ripening. Under normal conditions, if the tomato is left to ripen on the vine and develop full colour the softening process also occurs, simultaneously. Thus the fruits are easily bruised and damaged on shipment. By inhibiting the polygalacturonase by antisense genes the tomato can remain on the vine until mature and be transported in a firm solid state. The Flavr Savr tomato which has improved flavour and shelf life was engineered by Calgene in the USA where it is now marketed. Such techniques will now be used in a wide variety of soft fruits. Other studies are considering ethylene synthesis or inhibition as a means of controlling fruit and flavour maturation.

5. Improved seed products—proteins and fatty acids The plant products present in seeds, particularly amino acids and fatty acids, are essential for human health. The amino acids like lysine, methionine, tryptophan, etc. must be a part of diet. Corn seed proteins are lacking lysine while those of legumes are deficient in methionine and cysteine—sulphur-containing amino acids. With the advent of biotechnology it is possible to identify specific gene encoding for a protein that can be modified by inserting one or few selected additional codes responsible for desired amino acids. Such modified gene is further transferred to plant where it shows tissue specific expression in seeds with modified content of amino acids.

The genetic modification of oil-producing plants such as soyabean, rape and canola to produce a wide range of cooking oils, lubricants, cosmetics, detergents, etc. has also found wide interest. The gene expressing for the enzyme which catalyses desaturation in seed fatty acid biosynthesis has been identified and based on antisense RNA technology, it is possible to introduce this gene into plants so that it can express only in seeds modifying their fatty acid composition.

6. Plants as factories to produce novel biochemicals and vaccines As scientists discover how individual genes contribute to a particular end product from a plant, it will become increasingly possible not only to alter existing plant products through genetic modification, but also to develop entirely new plant products. Plants are potential factories for high value biochemicals like starch, sugar, lipids, proteins, and products like fine chemicals, perfumes and adhesive compounds as well as industrial lubricants, biodegradable plastic and even 'renewable' energy crops to replace fossil fuels. Similarly plants can be used to manufacture medicines by equipping them with the appropriate genetic material to produce antibodies, vaccines, painkillers and other medicinal proteins. One advantage of plant-based drug production is that unlike mammalian tissue culture, it is free from the risk of latent animal disease.

More recently, a team of scientists at the University of London, working with colleagues in Japan and Germany, produced genetically modified tomatoes containing three times the usual amount of vitamin A.

Besides fighting off cancer and heart disease, vitamin A is believed to help prevent macular degeneration, an eye condition that can lead to blindness. Similarly, rice can be converted into a life saving crop. It can be genetically modified to produce beta-carotene, which would prevent vitamin A deficiency that kills two million children each year and causes another 500,000 to go blind. This could be crossed with another variety with high levels of iron to combat anaemia.

Researchers at the Centre for Novel Agricultural Products at the University of York, UK are currently developing many novel products offering opportunities for commercial exploitation. A few of them are listed below:

a. A 'superglue' produced by tobacco plants with genes encoding for powerful adhesive proteins enables marine mussels to stick to rocks. It will be especially valuable as a biochemical glue for body repairs during surgery.

b. Transgenic plants, containing oil-encoding gene from marine algae, produce oil that has nutritional value similar to cod-liver oil.

c. Plant that will produce the antimalarial drug, Artemisilin.

d. Genetically engineered opium poppy to produce more powerful painkillers.

Plant genetic modification is an important tool in making edible vaccines. The conventional vaccines which were utilised for immunisation of animals or human beings normally made, while edible vaccines are produced from transgenic plants in which an orally active antigen of the target pathogen is expressed and accumulated, and then is fed raw to animals or human beings to bring about immunisation against the pathogen. Recently transgenic potatoes containing the edible vaccine provided with *E.coli* heat liable enterotoxin (LT)-B subunit. It is practically proved by the experiment that potato tubers containing LT-B were fed raw to mice and they showed immune response. The technology for producing plant-derived edible vaccines is cost effective and it can overcome the problems of storage and transportation, especially in developing countries like India.

The John Innes Centre, located on the Norwich Research Park, UK, is investigating whether plant viruses could be used to make novel vaccines for diseases such as HIV by transporting a small harmless fragment of the disease-causing virus into the plant, where it would be reproduced in large quantities.

Many scientists believe that genomics and gene discovery offer new opportunities for land use and that novel agriculture will one day bring new employment to rural areas.

4.5 ROLE OF ANTISENSE TECHNOLOGY IN TRANSGENESIS

The central dogma of molecular biology consists of the flow of genetic information from DNA via mRNA to protein.

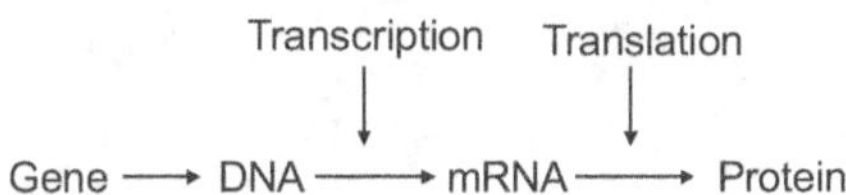

In the double-stranded DNA, the strand that codes the gene is called 'sense 'strand. Antisense RNA or DNA is a single-stranded nucleic acid, which is complementary to coding on 'sense' strand. It is also complementary to mRNA. The flow of genetic information, from DNA to mRNA in the form of codons then to protein, can be blocked by the introduction of RNA sequence complementary to the sequence of the target mRNA. Thus RNA duplex formed between mRNA and antisense RNA is either rapidly degraded or the mRNA is impaired in nuclear processing or it is blocked for translation. Antisense RNA for specific gene sequence can be generated by reversing the coding sequence of a gene under the control of promoter in normal orientation (figure 4.9).

The antisense RNA approach has been successfully applied for the following purposes.

1. Reducing polygalacturonase (PG) that degrades pectic polymers present in fruit cell wall. Transgenic tomatoes containing antisense gene construct for this enzyme, showed reduction in PG and slowed down the ripening and softening process and increased shelf life.

2. Reducing stearoyl-ACP (acyl carrier protein)- desaturase activity in seeds of transgenic *Brassica* species. This enzyme catalyses the desaturation step in seed fatty acid biosynthesis. Thus fatty

acid composition of vegetable oils present in seed can be modified.

3. Reducing chalcone synthase (ChS) activity in transgenic *Petunia*. ChS is a key enzyme of flavonoid biosynthesis. The flavonoids are necessary for normal development and functioning of pollen. It has been proposed to apply flavonol during pollination of ChS antisense male sterility (MS) lines to obtain 100% male sterile progeny.

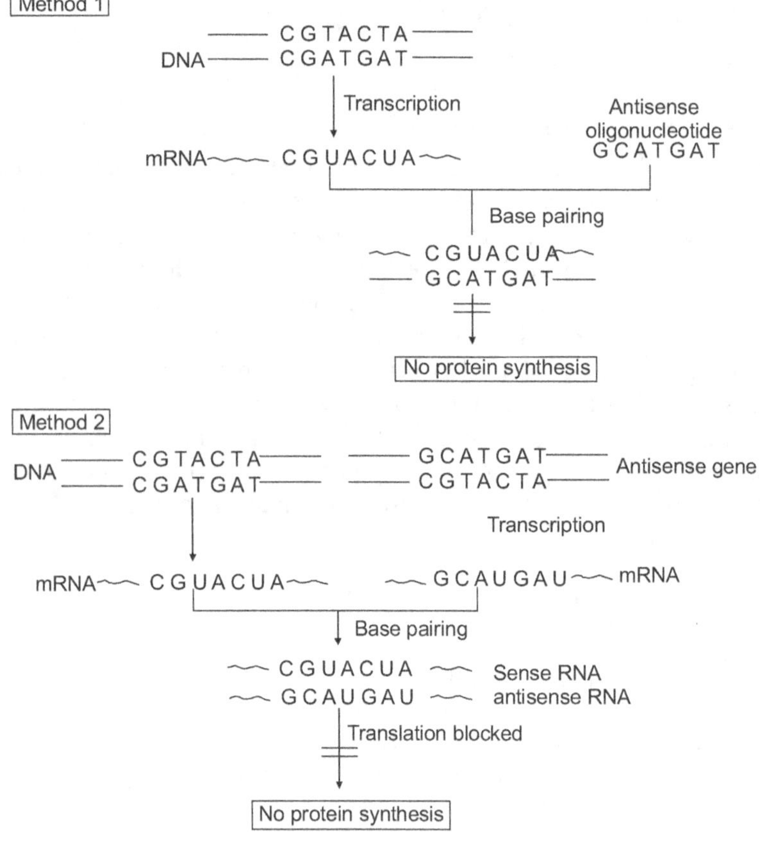

Figure 4.9   Two methods used to produce antisense effect by blocking the translation of mRNA

In addition to modification of plant properties, there are about 50 antisense pharmaceutical products in different stages of animal studies. The group of researchers at National Chemical Laboratory, Pune has synthesised a biocompatible fat-soluble compound that penetrates human cell barrier and delivers therapeutic DNA strands in the cell. These DNA fragments block the activity of genes of disease-producing viruses.

Key Concepts

- **Transfection** is the introduction of gene/s of interest into organisms to modify its characteristics for the welfare of human beings. The technique is also called transgenesis or gene transfer technology.

- With the advent of gene transfer technology it is possible to introduce a cloned gene into fertilised eggs and their successful implantation into surrogate mother to obtain the progeny carrying desired gene.

- The first successful transgenic mice were produced by introducing a gene for rat growth hormone (GH).

- Transgenic sheep achieved better growth, meat production and can also serve as bioreactors for synthesis of blood clotting factor IX and alpha-1-antitrypsin significantly involved in blood coagulation, and combating cystic fibrosis respectively.

- Transgenic pigs allow a ready supply of organs for transplantation.

- Transgenesis in plants is specifically aimed to improve resistance of plants to herbicides, insect pests, microbial diseases, and post-harvest features and to produce novel biochemicals and edible vaccines.

- Transgenic plants can be used as factories to produce novel biochemicals like starch, sugar, lipids, proteins, perfumes, adhesives, lubricants, biodegradable plastics and biomedical substances like antibodies, painkillers, medicinal proteins and edible vaccines, antimalarial drugs, beta-carotene, etc.

Review Questions

1. Define transfection. Describe various transfection methods for animal and plant cells.

2. What is Ti plasmid ? Explain its organisation with reference to T-DNA.

3. Define transgenic animal. What are the objectives of transgenesis in animals?

4. Describe the embryonic stem cell transfer technology for the production of knock out mice.

5. Comment on the following statements giving proper justification.

    a. Edible vaccines have prosperous future than conventional vaccines.

    b. Mice are the most suitable animals for transgenic production.

6. Describe the various applications of transgenesis in animals .

7. Plants have immense importance as bioreactors. Explain the statement with the help of suitable examples.

8. Describe important cases of gene transfer in some animal species with their significant applications.

9. What is Bt-endotoxin? Explain its mechanism in the development of insect resistance in transgenic plants.

10. Write short notes on:

a. Agroinfection	i. Knock out mouse
b. Plant viruses	j. Tracy—the transgenic ewe
c. Liposomes	k. Ideal herbicides
d. Particle gun	l. Cry proteins
e. Electroporation	m. Transgenic tomatoes
f. Microinjection	n. Edible vaccine
g. Super-mouse	o. Antisense gene
h. Inner cell mass	

5

BIOTECHNOLOGY, HUMAN DISEASES AND MEDICINE

5.1 INTRODUCTION

Human health is under the constant threat of various diseases caused by (i) invading microbes (e.g. AIDS, hepatitis, polio, leprosy, candidiasis) (ii) parasitic attack (e.g. sleeping sickness, filariasis, taeniasis, trichinosis, etc.) or (iii) imbalanced body's natural chemistry and genetic defects (e.g. cancer, heart disorders, dwarfism, diabetes, muscular dystrophy, sickle cell anemia, hemophilia, cystic fibrosis, etc.). Until this century, life for most people was harsh, lacking adequate nutrition, good sanitation and housing and, above all, longevity. But the contribution of biotechnology to the field of human health care is tremendous. Nowadays not only has the life style of people improved but also has their life expectancy increased from 35 to almost 80 years. These achievements are due to the spectacular role of biotechnology in the following areas:

i. Prevention of diseases with the help of therapeutic products like antibiotics, antibodies, vaccines, hormones, regulatory proteins, etc.

ii. Prenatal diagnosis of genetic diseases and genetic counselling.

iii. Immunodiagnostic tools and probes aiding disease identification.

iv. Correction of diseases either by enzyme or gene replacement therapy.

v. Improved contraceptives and vaccines to control fertility.

vi. Forensic medicine.

5.2 ANTIBIOTICS FOR PREVENTION OF MICROBIAL DISEASES

Microbiology (the science of study of microbes) has revealed that not all microbes are harmful. Humans play the role of host for literally millions of microbes and we would not survive long if most were not benign. With the advent of biotechnology, our relationship with microbes is entering a new phase—some microbes have been put to very positive medical uses—particularly in the production of antibiotics.

In 1928, the discovery made by Alexander Fleming that a fungus called *Penicillium notatum* could produce a compound selectively able to inactivate a wide range of bacteria without affecting the host, triggered the scientific studies that profoundly altered the relationship of humans to the controlling influence of bacterial diseases. In 1945, Guiseppe Brotzu, a Sardinian professor of bacteriology, found a microorganism in sea near a sewage outfall. This fungus, a species of *Cephalosporium*, produced a substance that killed a wide range of bacteria. Brotzu did not have facilities to analyse this substance, so he sent the organism to Oxford. There, Guy Newton and Edward Abraham discovered that the fungus manufactured a novel type of Penicillin, which they named Penicillin N. Then in 1953, they made another discovery of much greater importance; this organism also manufactured another antibiotic, which they called Cephalosporin C.

Antibiotics are antimicrobial compounds produced by living microorganisms, and are therapeutically and sometimes prophylactically used in the control of infectious diseases like pneumonia, tuberculosis, cholera, leprosy, typhoid, fungal diseases, etc.

Cephalosporin C has the immediate advantage of being able to kill bacteria that have become resistant to penicillin. The phenomenon of resistance to antibiotics has proved to be the major driving force behind the quest for new antibiotics. Today there are about 4000 antibiotics that have been isolated (but only 50 have achieved wide usage) to kill or disable other microbes (table 5.1). The large number of antibiotics that are not used in medicine are rejected for various reasons, some produce too many harmful side effects, some are too expensive to manufacture on a large scale, and some simply cannot do a specific job.

Antibiotics have been extensively used in medicine since 1945, with the arrival of penicillin. Antibiotics are now widely used in human and veterinary medicine and to some extent in animal farming where some antibiotics have been shown to increase the weight of livestock and poultry. Antibiotics can also be used to a limited extent to control plant diseases and as insecticides.

Both penicillin and cephalosporin are members of the group of antibiotics known as beta-lactams named after the characteristic type of chemical ring structure that they possess. They operate by preventing certain bacteria from building proper cell walls. On the other hand, streptomycin belongs to the group of antibiotics known as aminoglycosides; they prevent certain bacteria from manufacturing their proteins by disrupting its ribosomes. Chloramphenicol blocks peptidyl transfer while erythromycin, an antibiotic, blocks chain extension soon after the initiation of protein synthesis in bacteria.

Table 5.1   Some commercially important antibiotics

Antibiotic compound	Source	Application
Actinomycin D	*Streptomyces* sp.	Antitumour
Amikacin	*Streptomyces* sp.	Antibacterial
Ampicillin	*Streptomyces* sp.	Antibacterial
Bacitracin	*Bacillus* sp.	Antibacterial
Cephadroxil	*Streptomyces* sp.	Antibacterial
Cephalosporin	*Acremonium* sp.	Antibacterial
Chloramphenicol	*Cephalosporium* sp.	Antibacterial
Ciprofloxacin	*Streptomyces* sp.	Antibacterial
Erythromycin	*Streptomyces* sp.	Antiprotozoan
Fluconazole	*Penicillium* sp.	Antifungal
Daunorubicin	*Streptomyces* sp.	Antibacterial
Fumagillin	*Aspergillus* sp.	Amoebicidal
Gentamycin	*Streptomyces* sp.	Antibacterial
Griseofulvin	*Penicillium* sp.	Antifungal
Mitomycin C	*Streptomyces* sp.	Antitumour
Natamycin	*Streptomyces* sp.	Food preservative
Penicillin G	*Penicillin* sp.	Antibacterial
Rifamycin	*Nocardin* sp.	Antituberculosis
Streptomycin	*Streptomyces* sp.	Antibacterial
Tetracycline	*Streptomyces* sp.	Antibacterial
Tobramycin	*Streptomyces* sp.	Antibacterial

The production of antibiotics has undoubtedly been a highly profitable part of the pharmaceutical industry in the developed nations. World market for antibiotics is worth over USD 10 billion per year and is the most valuable segment of the total pharmaceutical market (about USD 200 billion). The improvements in production technologies (such as protoplast fusion and gene transfer techniques) are leading to develop new strains of bacteria with high productivity, improved stability and possible new products with decreased costs in production. The advances in biotechnology may make it possible to follow a more enlightened pathway to develop the low cost antibiotics to fight the massive specific disease problems of developing nations.

A disturbing observation has been the gradual evolution of drug resistance in many bacteria. The possibility of this acquired resistance, which is transmitted from one bacterium to another is now a fact. For example, gonorrhoea (a veneral disease) resistant to treatment with Penicillin is now prevalent in 19 countries. It is known that the resistance factor is located on plasmids within the bacterium that are easily transmitted among organisms. Antibiotic resistance in many well-known diseases is steadily increasing which is of serious concern in our society.

New antibiotics   When bacteria become resistant to antibiotics, the antibiotics have to be redesigned. The problem is that the bacteria may then become resistant to the altered antibiotics. Research at the University of Cambridge is focusing on creating an entirely new generation of antibiotics by using DNA sequence analysis to mix and match the ingredients of existing ones, while scientists at the Sanger institute, UK, are studying the *Streptomyces* genome with a view to making antibiotics more effective and easier to produce.

5.3 ANTIBODIES IN IMMUNITY

Antibodies are soluble proteins produced by human and animal systems as a part of immune response against invading pathogens. The interaction of antibodies with the surface of a virus or bacterial cell neutralises the pathogen's ability to infect a host cell and facilitates the pathogen's ingestion and destruction by wandering phagocytes. The immune system is capable of producing millions of different antibody molecules that are capable of recognising and binding to virtually any foreign substance, or antigen to which the body might be exposed.

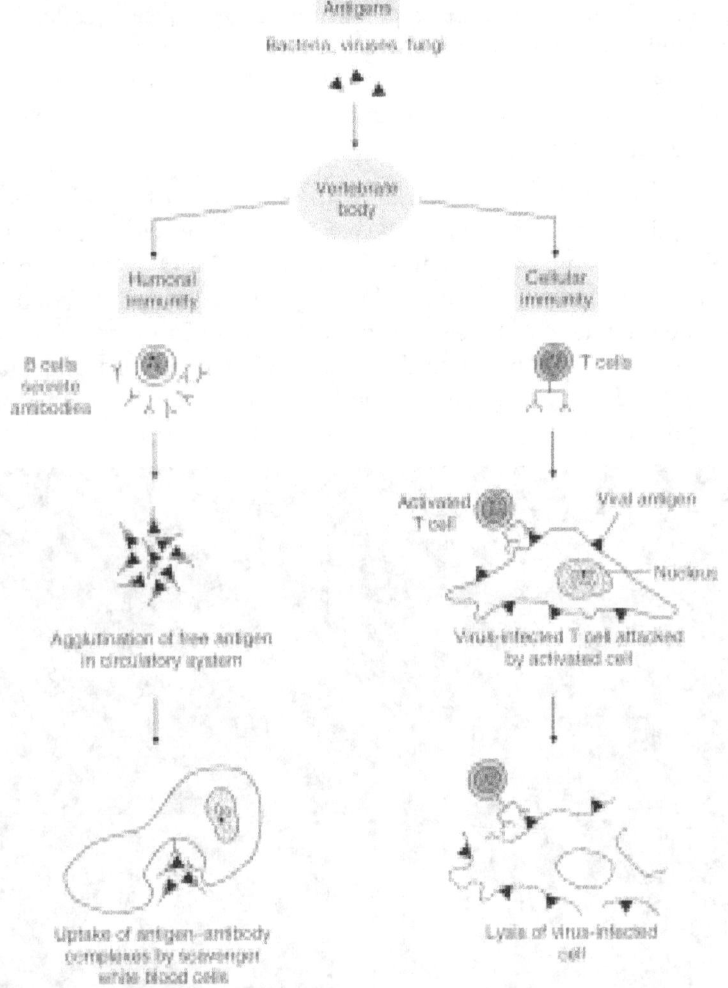

Figure 5.1   Diagrammatic representation of immune response to antigen
             in the body of vertebrates

In general, antigens are proteins, or proteins combined with other substances that reside on the surface of the invading microorganism and trigger the body's defence against it. In this way antibodies are the essence of immunity against diseases. There are two forms of immune response in the human body. The humoral response involves the cellular secretion of antibodies into the serum of blood, in response to specific foreign substances and they are partially responsible for their elimination. The other type of immune response is called cell-mediated response that involves cells of immune system directly attacking and destroying tissues and/or cells recognised as foreign. Unlike the humoral response, the cellular immunity produces no antibody (figure 5.1).

Antibodies or immunoglobulins are built of two types of polypeptides chains, the larger heavy chains (Mol. wt. of 50 KD to 70 KD) and smaller light chains (Mol. wt. of 23 KD). The two types of chains are linked to one another in pairs by disulphide bonds to give the schematic Y-shaped structure as shown in figure 5.2. Each chain also has regularly spaced intra chain S–S bonds. Carbohydrate moieties (CHO) are attached to heavy chains, making the antibody molecules glycoproteins.

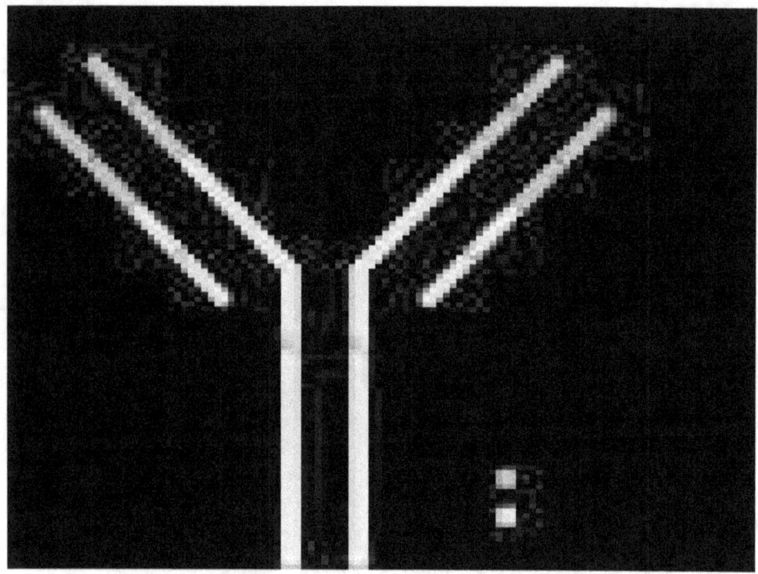

Figure 5.2   A molecular model of a typical human antibody, composed of four polypeptide chains joined by disulphide bonds

Five different classes of immunoglobulins (IgA, IgD, IgE, IgG and IgM) have been identified. The different immunoglobulins appear at different times after exposure to a foreign substance or in different body fluids (table 5.2).

Table 5.2   Classes of Human Immunoglobulins

Class	Molecular Weight in KD	Properties
IgA	360–720	Present in tears, nasal mucous, breast milk, and intestinal secretion.
IgD	160	Present in various cell plasma membranes, function unknown.
IgE	190	Binds to most cells, releasing histamine responsible for allergic reactions.
IgG	150	Primary blood-borne soluble antibodies; crosses placenta.
IgM	950	Present in B cell plasma membranes, mediates initial immune response, activates bacteria-killing complement.

Currently, antibodies are obtained from immunised animals, which is a tedious and time-consuming operation. At the end of various extraction and purification stages the antibodies become weakly specific, available only in small batches and of variable activity. Attempts to culture antibody-secreting cells have been unsuccessful, since such cells neither survive long enough nor produce enough antibodies in culture to become worthwhile sources of antibodies. Furthermore, such systems normally produce mixtures of different antibodies (polyclonal antibodies).

5.4 MONOCLONAL ANTIBODIES IN VARIETY OF DIAGNOSIS

A significant development in medical biotechnology has been the ability to produce monoclonal antibodies. When antibody-producing cells are immortalised and stabilised the secreted antibodies will always be the same from that particular cell line. Monoclonal antibodies are usually produced from hybridoma clones (as described in Section 2.10) or by various forms of fermentation technology. Monoclonal antibodies are now finding wide applications in diagnostic techniques requiring highly specific reagents for the detection and measurement of soluble proteins and cell surface markers in blood transfusions, haematology, histology, microbiology and clinical chemistry, as well as in other non-medical areas. Monoclonal antibodies are employed for:

i. Early and accurate diagnosis of cancer and subsequent therapy.

ii. Classification of blood groups (ABO, Rh, etc.).

iii. Diagnosis of pregnancy.

iv. Diagnosis of sexually transmitted diseases.

v. Prevention of immune rejection of organ transplant.

vi. Monoclonal antibodies could also be used in the preparation of very specific vaccines, particularly against viral strains and against other parasites.

vii. Clear and specific detection of various pathogens.

viii. Purification of industrial products.

ix. Detection of trace molecules in food and agriculture industry.

On a commercial scale monoclonal antibodies are being produced in 100-litre airlift fermenters by encapsulation and in perfusion chambers using lymph from live cattle. Commercially, monoclonal antibodies have been one of the most rewarding areas of modern biotechnology.

5.5 VACCINES FOR IMMUNISATION

Vaccines are biological preparations utilised for prevention of diseases. In 1967 over 10 million people were infected with small pox and the disease was endemic in more than thirty countries. Today, this dreaded disease has been wiped off the map. The World Health Organization's mass vaccination programmes against small pox is arguably the greatest triumph of modern medicine, and its success triggered the development of effective vaccines against viral diseases. Most vaccines against viral infections such as polio, small pox, yellow fever, rabies, rubella (German measles) contain either whole or killed viruses that are no longer capable of causing serious infections. Injection of these harmless viruses (vaccination or immunisation) tricks the body's immune system to produce specific antibodies or immune cells that lie in wait, ready to attack the authentic disease-causing virus should it happen to enter the body. One of the advantages of using live, attenuated viruses as part of a vaccine is that it stimulates both the humoral and cell-mediated pathways of the immune system.

Hepatitis B (serum Hepatitis) is a dreadful viral disease. The virus that causes this disease has been identified and is responsible for thousands of serious liver infections and roughly 100 deaths a year in Britain, while in the US, the Centre for Disease Control (CDC) estimates that there may be as many as 150,000 cases a year, resulting in about 100 deaths. In Africa and Asia, the disease is even more common, where, as much as 10% of the population suffers from hepatitis B. Furthermore, hepatitis is thought to be a major cause of liver cancer. Considering the need of time, genetic engineers prepared the vaccine against hepatitis B using recombinant DNA technology. This vaccine is based on recombinant proteins (means proteins produced by recombinant DNA technology), since viruses are amazingly simple structures consisting of just a small piece of genetic material encased in a coat of protein. The proteins found in virus coat are generally immunogenic. The genes encoding such proteins are identified and isolated from the virus and then introduced into the bacterium, *E.coli* where the genes are expressed to produce proteins on a large scale. These genetically engineered proteins are then purified and mixed with suitable stabilisers and used for immunisation. This is the method for preparation of hepatitis B vaccine (H-B-Vax), which is commercially used, for controlling the disease.

Human Immunodeficiency Virus (HIV) causes Acquired Immunodeficiency Syndrome (AIDS) that is another life-threatening disease. It was first discovered in 1981, at that time, it was restricted almost entirely to a few specific groups, notably male homosexuals, intravenous drug users, haemophiliacs and Haitians. Today, it has spread like a fire all over the world. It is estimated that by the year 2005 approximately 60 million people will be infected with HIV, over 90 percent of these individuals will be living in the poor third world countries. Taking these facts into consideration, a number of companies, in collaboration with various government agencies around the world, are trying to produce vaccines composed of HIV envelope proteins. There are two problems in the development of such vaccine. Firstly, unlike most viruses, HIV is capable of rapidly mutating, causing changes in the structure of its envelope proteins.

Therefore, the antibodies produced against envelope protein that is present on the viruses of cultured cell line from which it was derived are not present on the viruses residing in the infected persons. These results caused many researchers (who are testing the efficacy of the vaccine) to become doubtful whether the vaccine would prove effective in preventing HIV infection in the population at large. The second problem is in enlisting the volunteers for the trial. In addition to a significant number of enlistees being given a placebo vaccine, those that received the vaccine would test HIV-positive from then on. This is because the test to determine HIV status depends on the presence of the same antibodies that the vaccine is designed to induce. The worst thing that happened during testing was that at least 10 individuals who had participated in earlier trials that were designed to determine the safety and immunological potency of the vaccine had become infected with the virus. These findings decreased the tempo of research groups belonging to Genentech and Chiron/Ciba Geigy, two US-based companies who were claiming for production of AIDS vaccine. The researchers remain pessimistic about the likelihood of developing an effective vaccine against AIDS. An alternative approach to prevent and treat AIDS involving the potential use of ribozyme is discussed in Chapter 6.

The group of researchers belonging to the John Inne Centre, UK is looking for an alternative way of producing the vaccines from the plants. They are investigating whether plant viruses could be used to make novel vaccines for HIV by transporting a small harmless fragment of the disease-causing virus into the plant for their production in large quantities. This may be an efficient and effective method of production of low-cost vaccines.

Acambis plc, the company which has operations in UK and USA, developed and manufactured vaccines against West Nile, Japanese encephalitis, dengue fever, typhoid and travellers' diarrhoea. In addition to this the US government awarded a contract to this company in September 2000 and November 2001 to develop and stockpile a new vaccine against smallpox to counter the threat of smallpox virus being used as a bioterrorist weapon.

Similar efforts are made by Powderject Pharmaceuticals Plc, based in England and USA, this company has become a world leader in the field of particle-mediated injection of medicines, developing novel products based on the use of its proprietary powder injection systems. This needle-free system painlessly delivers medicines and vaccines through the skin in a dry powder form with the help of a small hand-held device. The company is a world leader in developing five new powderject DNA vaccines that entered clinical trials in 2003. These vaccines are:

a. Powder injection flu DNA vaccine

b. Powder injection herpes simplex virus (HSV) vaccine

c. Powder injection cancer therapeutic vaccine

d. Powder injection defence vaccine for US military

e. Powder injection HIV DNA vaccine

5.6 HORMONES FOR CORRECTING BODY'S CHEMISTRY

Hormones are the body's chemical messengers. They are complex organic molecules with varying chemical composition, secreted by ductless or endocrine glands. Our body contains a dazzling array of different hormones, each has the vital task of coordinating the activities of individual cells and tissues. Without hormones whole organism cannot function properly. Hormones transfer all sorts of information from one group of cells to another distant tissues. A defect in one of the hormone systems can produce serious illness that can sometimes be corrected by supplying the patient with the right amount of a particular hormone; biotechnology can play an important role in manufacturing the hormone. Insulin and human growth hormone are two such examples of hormones manufactured by genetic engineering.

5.6.1 Insulin

Insulin is a hypoglycaemic hormone secreted by the beta cells of the islets of Langerhans of the pancreas. The deficiency of insulin causes a disorder known as diabetes mellitus. Until the advent of genetic engineering all the insulin given to diabetics was extracted from the pancreas of cattle or pig. The insulin molecule is made up of 51 amino acids; pig insulin has a different amino acid at one point as compared to the human version, whereas cattle insulin differs at three points. These small variations mean that some patients eventually suffer allergic reactions. To avoid this, and considering the ever-increasing number of diabetic patients, biotechnology made it possible to produce safe, efficient, low-cost human insulin on a large scale from genetically engineered bacteria. The process of manufacture of humulin from genetically modified microbes is discussed in Chapter 3 (Section 3.7). Recombinant human insulin appears to have no side effects and is increasingly having the largest market share of sale. Production is unlimited and free from market shortage of animals and all other problems associated with previous production methods.

5.6.2 Human Growth Hormone (GH)

Somatostatin or growth hormone is secreted by anterior pituitary. It stimulates the growth and development of tissues, bones and muscles of the body. In children, hyposecretion of growth hormone produces dwarfism while in adults low secretion of GH leads to Simmond's disease (dry and wrinkled skin with premature ageing and degenerated sex organs). If the lack of GH is diagnosed early enough, such children can be given extra GH to maintain their growth. By cloning the human gene for GH into a bacterium, the hormone can be produced from transgenic bacterial fermentation. A single 500-litre vessel filled with genetically engineered bacteria can provide GH equivalent to no less than 35,000 human pituitary glands. In 1981, in the UK, 800 children could have benefited from treatment. In the US, about 2600 children are receiving GH. One child in 5000 suffers from hypopituitary dwarfism and the easy availability of this biopharmaceutical will be of immense benefit to these suffering children. It is also evident that the GH can increase muscle formation in normal individuals and is now being exploited by some athletes.

Eli Lilly and Company Ltd., one of the UK's top pharmaceutical company manufactured the human growth hormone, Humatrope by recombinant DNA technology. The biological effects of Humatrope are similar to the human growth hormone made in the pituitary gland.

Biotechnologists are trying to manufacture equally vital hormones including calcitonin, cholecystokinin, vasopressin, parathyroid hormone, nerve growth factor, adrenocorticotropic hormone, and erythropoietin from genetically engineered microbes.

> ## Key Concepts
>
> ❑ Biotechnology plays an indispensable role in diagnosis and cure of diseases.
>
> ❑ Antibiotics are antimicrobial compounds produced by living microorganisms and are effectively used in control of various infectious diseases.
>
> ❑ Antibodies are immunoglobulins produced by human and animal systems in response to antigen present on viruses and bacteria that invade the body.
>
> ❑ Human body shows two types of immune responses specific to foreign protein—the humoral response involves production of antibodies and cell mediated response involves cells of immune system that directly destroy the antigen.
>
> ❑ Hormones are chemical messengers secreted by endocrine glands. Their inadequate quantities results in metabolic disorders.
>
> ❑ Pharmaceutical companies are engaged in production of hormones like Humulin and Humatrope by techniques that are based on recombinant DNA technology.

5.7 INTERFERONS AND CANCER CURE

Interferons are glycoprotein molecules produced by a variety of cell types and provide protection to other healthy cells from virus attack. In 1957 two British researchers, Isaacs and Lindenmann discovered substances produced within the body that could act against viruses by making cells resistant to virus infection. These substances, which interfered viral infection, are called interferons.

Interferons are classified into 3 major types as follows:

i. Interferon-α produced by leucocytes or white blood cells.

ii. Interferon-β produced by fibroblasts.

iii. Interferon-γ produced by stimulated T- lymphocyte cells hence also called immune interferons.

Much of the current interest in interferons has arisen from their ability to inhibit cancer in experimental animals. The mode of action of interferons can be summarised as follows.

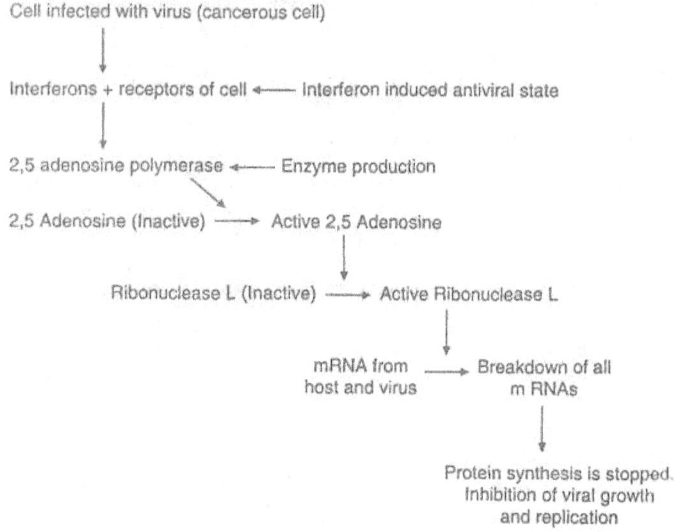

Interferons can also stimulate the body's natural immune defence against the cancer cell. They have been approved for use in treatment of cancer and other viral diseases e.g. Hepatitis B and AIDS.

Currently interferons can be obtained from two sources. The first is human diploid fibroblasts growing attached to a suitable surface and the second source is recombinant bacteria having human fibroblast interferon gene into its plasmid. Interferons synthesised in this manner can be extracted and purified.

In addition to interferons, white blood cells produce lymphokines as a part of body's natural defenses against cancer. Lymphokines are considered to be crucially important to immune reaction. Interleukin-2 and lymphotoxin are two such proteins having the capability of fighting cancer. They have great potential in cancer therapy.

5.8 DNA PROBES AND DISEASE DIAGNOSIS

Quick and accurate diagnosis of disease and its causative agent is the first and crucial step in management and cure of disease. Microscopic examination or culture of the specimen can provide the information regarding the pathogen but a recent technique that has grown out of genetic engineering research proves more powerful. Short pieces of DNA called probe designed stick to the DNA of a particular bacterium present in the patient's blood provide a diagnostic tool. The implications of DNA probes are profound. They not only assist in identification of microbial infections but also indicate genetic defects and contaminations in donated blood, help in matching organs for transplantation and even assist seed suppliers to test the quality of their seeds.

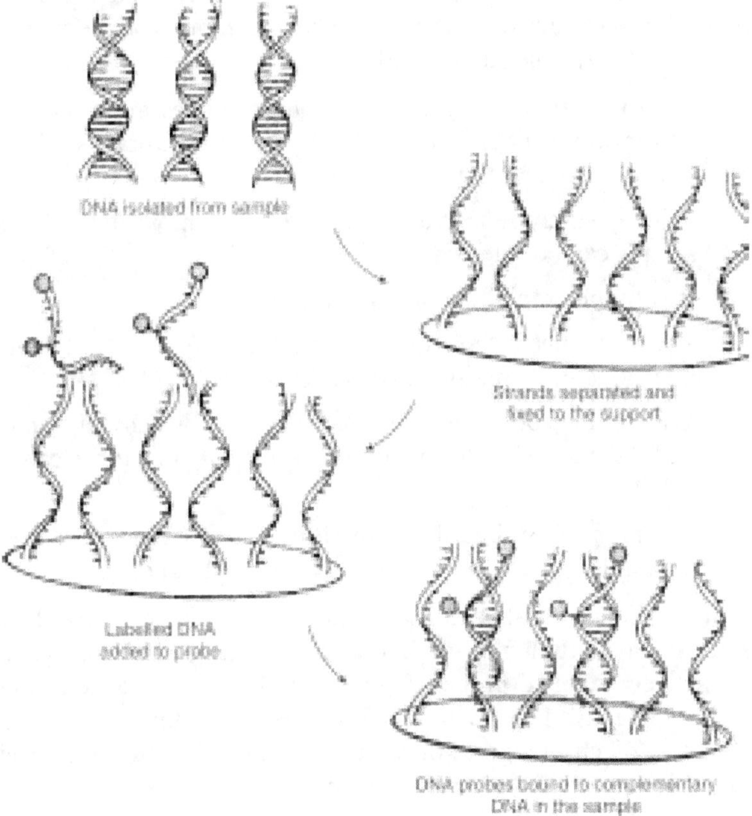

Figure 5.3   Mode of action of DNA probe in diagnosis of complementary
              DNA sequences in the sample DNA

Figure 5.3 shows the mode of action of DNA probes. The technique depends on the fact that two strands of DNA that have complementary sequence of bases will stick together. To check if a sample contains a particular type of DNA (say from a microbe) a biotechnologist will mix a DNA probe into the sample. The probe has been designed so that it is complementary to some part of the DNA that is being sought out. If the probe finds it's opposite strand with complementary base sequences, the two link together. The probes have previously been labelled with a radioactive compound. The probes attach themselves to sample DNA and any that remain free are then washed away. The amount of the labelled probe still fixed to the support via the sample DNA is measured. If the sample DNA is not the type the probe has been designed to latch onto, then no probes will remain on the support and the test gives a negative result.

DNA probes are claimed to be a thousand times more sensitive than conventional diagnostic methods. Some advantages of this diagnostic tool are as follows.

1. It is extremely sensitive and even a single molecule in a test sample can be detected.

2. It is simple, rapid and highly specific.

3. There is no need of microbial culture; therefore risk of accidental infections in laboratory workers is avoided.

4. With the help of Polymerase Chain Reaction, which can amplify DNA sequences upto million fold sensitivity of probes can be enhanced.

5. Number of probes for detection of protozoan parasites, helminthes, bacteria and virus are easily obtained.

Non-radioactive labelled probes are also available to avoid the health hazards that may arise due to handling and disposal of radioactively labelled probes. Signal amplifications and detections of non-radioactive label is achieved by specific binding and enzyme conversion of chromogenic or bioluminescent substrate to give out detectable signal (colour change or luminescence).

5.9 ELISA AND IMMUNO-PCR AS IMMUNODIAGNOSTIC TOOLS

Enzyme-linked immunosorbent assay (ELISA) is the most rapid and highly efficient method employed for diagnostic purpose. It is a sensitive serological test used for detection and quantification of

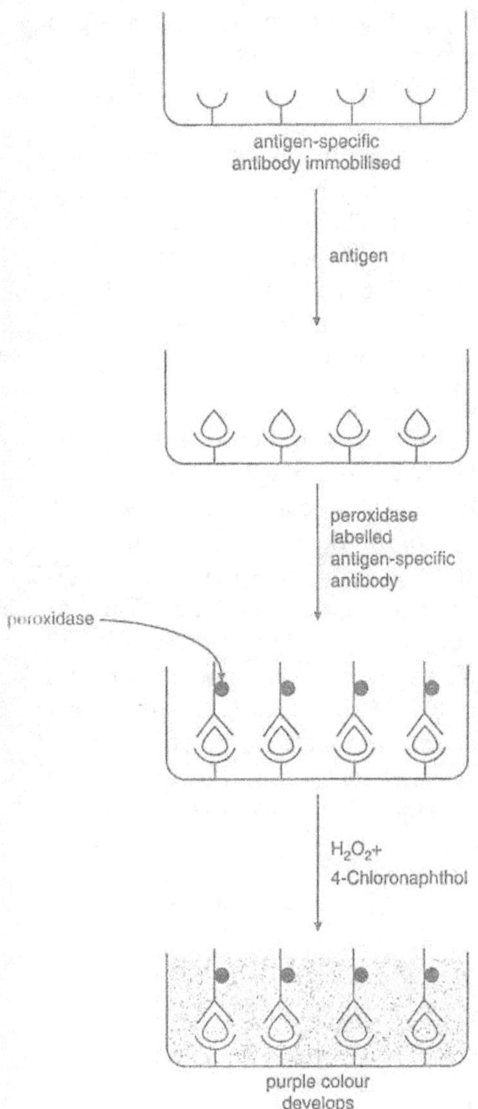

Figure 5.4   Diagrammatic representation of double antibody sandwich (DAS) ELISA

viruses, proteins and small molecules such as hormones, formed on a microtiter plate. Many samples can be tested at the same time, both rapidly and economically. The most widely used form of ELISA test uses the double antibody sandwich (DAS) technique. In this technique, an unlabelled antibody (Ab) specific to the antigen (Ag) of interest is immobilised in a microtiter well. The antigen is added and allowed to react with the immobilised antibody to form Ag-Ab complex. A second antibody specific to the antigen, and labelled with an enzyme is now added which reacts with the Ag-Ab complex. The unreacted antibodies are washed out, enzyme substrate, etc. are added and the colour change is measured. The most commonly used enzyme for labelling of antibodies is peroxidase, its substrate hydrogen peroxide (H_2O_2) is converted into H_2O and O_2 in the presence of electron donors like 4-chloronaphthol which is oxidised in the reaction to produce purple colour, its intensity provides quantitative assay of the antigen of interest (figure 5.4).

Recently a more sensitive technique called immuno-PCR is developed that can detect a single antigen present in the sample. It is an extremely versatile antigen detection system in which PCR is used to amplify a segment of marker DNA that has been attached specifically to the antigen–antibody complex. Immuno-PCR includes a molecular linker having biospecific binding affinity for antibody, and DNA is used to attach a marker DNA molecule to antigen-antibody complex. With the help of a suitable primer, marker DNA is amplified in PCR and the products are analysed by gel electrophoresis. The most appropriate molecular linker is streptavidin-protein A that binds to antibody on one hand and to biotinylated marker DNA on the other hand. The overall sequences can be summarised as follows.

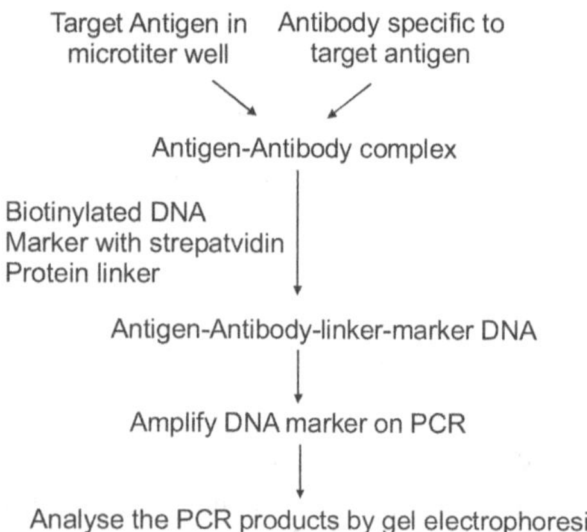

Immuno-PCR is employed in clinical diagnosis since it is more sensitive than ELISA. It is an extremely useful technique for early detection of pathological conditions as it can detect rare antigens at the single cell level.

In addition to the above-mentioned immunological diagnostic tools, **autoantibodies** can also be used in clinical diagnosis of **autoimmune diseases**. We know that the immune system normally produces antibodies against foreign proteins but not against the native proteins of the body. In other words the immune system can distinguish between 'self' and 'nonself'. However, in rare cases, individuals begin to produce antibodies against their own antigens. These antibodies are called autoantibodies and the diseases resulting from their presence are known as autoimmune diseases. Among these diseases are myasthenia gravis (antibodies against one's own muscle cell acetylcholine receptors), systemic lupus erythromatosus (antibodies against one's own nuclear DNA), paroxysmal cold haemoglobinuria (antibodies against one's own red blood cells) and primary biliary cirrhosis (antibodies against one's own mitochondria).

5.10 PRENATAL DIAGNOSIS OF GENETIC DISEASES

Diagnosis of chromosomal aberrations (like aneuploidy, polyploidy, deletion, translocation, etc.) as well as sex prediction is possible in prenatal conditions i.e. before the birth of child. Removing amniotic fluid by puncturing the mother's abdomen, a procedure called amniocentesis, which does not provide any harm to the foetus. The amniotic fluid contains cells shed from the foetus and these cells can be cultured (figure 5.5).

Amniocentesis performed at mid-trimester of gestation, involves removing amniotic fluid with the help of hypodermic needle from the amnion surrounding the foetus in the uterus, culturing the foetal cells present in the fluid and further screening of cultured cells for the following:

i. Karyotype to detect chromosomal abnormalities such as Down's syndrome, cry-du-chat syndrome, etc.

ii. Detection of errors in metabolism by studying enzyme production (diseases like cystic fibrosis, Gaucher's disease, phenylketonuria, etc.).

iii. Detection of defects in the neural tube of the embryo by analysing α-feto protein. Other genetic abnormalities such as those involving the haemoglobin molecule (i.e. sickle cell anaemia and β-thalassemia) could be revealed only by removing foetal blood but recently a combination of amniocentesis with analysis of restriction enzyme fragments helps scientists to do prenatal diagnosis of these genetic defects.

At present nearly 200 inborn disorders can be detected *in utero*. In addition to amniocentesis, more recently a new technique for prenatal diagnosis has been developed which is based on obtaining a biopsy of the chorionic villi present at an early stage in embryonic development. This technique allows prenatal testing during the first 3 months of pregnancy, a time at which amniocentesis is not possible. The tissue obtained by a small catheter is made of rapidly dividing embryo cells that can be diagnosed readily for chromosomal and biochemical disorders.

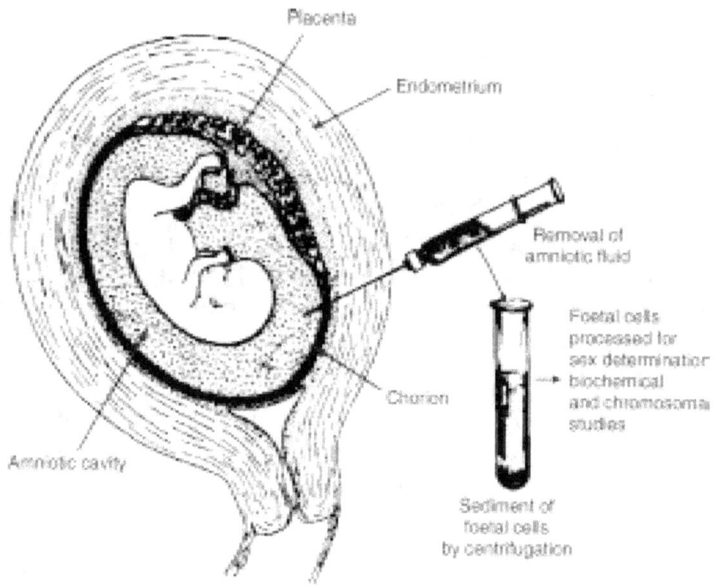

Figure 5.5   Steps involved in amniocentesis technique

5.11 GENETIC COUNSELLING

It is the advice rendered by a medical professional or geneticist for the people who are known to, or suspected to, carry certain genetic defects. Counselling is designed to inform people about the risks of having affected children and the likely consequences, the decision is then left in the hands of the individuals concerned, it cannot be made by the counsellor.

A well-qualified person in this field can advice voluntary restriction of child bearing by a couple that carries a serious hereditary defect through proper counselling. For instance if an individual is aware of his father suffering from the serious genetic ailment he himself would not want his children to suffer from the same fate, rather he would not produce any children. The obvious benefit of genetic counselling is that the couple will voluntarily restrict themselves to give birth to children with genetic disorders. Thus the method of counselling can do great benefit to human society. Such genetic counselling units exist in many western countries including USA. Unfortunately in India, counselling centres are less and even if they exist they are not popular. And also in countries like our own, where people are generally poor, genetic counselling should be provided free and must be funded by governmental or non-governmental organisations. People are generally unwilling to pay for such counselling, although they will pay for services of a doctor for diagnosis and treatment of the disease.

5.12 ENZYME-REPLACEMENT THERAPY

Three percent of humans have hereditary diseases that are transmitted in Mendelian fashion. Every one of us carries about half a dozen defective genes. We remain unaware of this fact unless we, or one of our close relatives, are among the millions of people who are suffering from a genetic disease. Gaucher's disease is one of the ten known hereditary lipid storage diseases, most of which cause extensive damage to the nervous system. Children suffering from this disease lack the enzyme that normally breaks down one of many forms of lipids in the body, glucocerebroside. This material accumulates in liver, spleen and bones, producing swelling, damage to nerves and sometimes, death before the age of two. If the right enzyme could be delivered to the cells, then the symptoms of Gaucher's disease could be relieved. Phenylketonuria (PKU) is another inborn error of metabolism arising due to lack of the enzyme phenylalanine hydroxylase that converts phenyalanine into tyrosine. The patients of PKU, are physically and mentally retarded with increased levels of phenylalanine in blood. At present the enzymes needed to combat Gaucher's disease and PKU are purified from human urine and placenta where they are present in very small quantities. It is time-consuming and expensive to collect sufficient enzyme for clinical trials, let alone the full-scale therapy in many patients. Biotechnologists are trying to solve this problem by making bacteria to manufacture the specific enzyme on a large scale.

The specific biochemical disorders, usually arising due to lack of a particular enzyme are known as metabolic diseases. In humans about 200 metabolic disorders have been identified, some of them can be treated by enzyme-replacement therapy taking the aid of genetically modified microbes.

5.13 GENE-REPLACEMENT THERAPY

There are more than 5000 different human genetic diseases known to be caused by single gene defects e.g. sickle cell anaemia, thalassemia, Tay-sach's disease, cystic fibrosis, Huntington's chorea, haemophilia, alkaptonuria, albinism, etc.

Even the disorders related to heart and cancer seem to have genetic components. The mutation in oncogene and tumour suppressor gene can transform the normal cells into malignant cells. In 2002, Professor Mike Stratton and his team of researchers working on the cancer genome project at Sanger institute, UK, identified a major genetic change that leads to 70 percent of malignant melanoma (a potentially lethal form of skin cancer), and a smaller proportion of other cancer types. The cancer genome project uses high-throughput mutation detection system to identify the oncogene—the mutated gene that causes cells to become cancerous.

Since inherited diseases arise due to mutation in a single gene, the ultimate cure would be to provide the patient with the correct gene that will help the body to make the required enzyme or other proteins. The gene replacement therapy is one of the most radical notions ever put forward in medicine, to eliminate the root causes of the disease.

The first attempt of gene replacement in humans were made in 1980 by Martin Cline and his colleagues from the University of Los Angeles to cure patients suffering from thalassemia, a severe and fatal blood disease caused by defects in the gene coding for globin. Red blood cells are rich in haemoglobin and are derived from bone marrow. The researchers removed some bone marrow cells from the patients and tried to introduce a normal globin gene into them (using the vectors analogous to the plasmids employed in genetic engineering of microbes) in the hope that this would take over the function of their own defective genes, allowing the patients to make normal haemoglobin. As globin genes were the first human genes to be cloned, there was no difficulty in obtaining normal genes for insertion into the marrow cells. After these cells had been genetically manipulated, they were replaced into the patient's bones. These operations however failed possibly because the genes entered the wrong part of the patient's chromosome and did not transcribe into mRNA, therefore globin synthesis could not occur.

In September 1990, one such attempt of gene therapy was made on a human patient (a girl) for severe combined immuno deficiency (SCID) produced by adenosine deaminase (ADA) deficiency at the National Institute of Health in Bethesda, USA, by Michael B. and W.F. Andresow. Currently, there are more than 20 companies all over the world working in this area of development. Gene therapy is the process by which a patient is cured by altering his or her genotype. This approach can be classified into two types:

5.13.1 Germ Line Gene Therapy

In this method healthy genes can be introduced into germ cells like sperms, eggs, early embryos. Though it is highly effective in counteracting the genetic disorders, it is not encouraged for application in human beings due to a variety of technical and ethical reasons.

5.13.2 Somatic Cell Gene Therapy

In this type the gene is introduced only in somatic cells like bone marrow cells, hepatic cells, fibroblasts endothelium and pulmonary epithelial cells, central nervous system, endocrine cells and smooth muscle cells of blood vessel walls. Modification of somatic cells only affects the person being treated and the modified chromosomes cannot be passed on the future generations. Somatic cell gene therapy is the only feasible option and the clinical trials have already employed for the treatment of cancer and blood disorders including SCID, Gaucher's disease, familial hypercholesterolemia, haemophilia, phenylketonuria, cystic fibrosis, sickle cell anaemia, Duchenne muscular dystrophy, emphysema, thalassemia etc. Somatic cell gene therapy is divided into two groups on the basis of the end result of the process.

    i. Gene Augmentation Therapy (GAT)

    ii. Targeted Gene Transfer

i. Gene augmentation therapy In this method the functional gene is introduced into the somatic cell in addition to the defective gene endogenous to the cell. Thus the modified cells contain both endogenous defective gene as well as normal introduced gene. There are different viral vectors used to introduce genes into target/stem cells cultured *in vitro*. Physical method of gene transfer like Ca^{++} phosphate precipitation, particle gun, electroporation, etc. are generally used to deliver the genes into target cells. Another approach to transfer the gene is the direct injection of DNA into specific tissues like liver in the form of a protein complex or it can be injected into muscle or skin in the form of naked DNA. The results of this approach are encouraging since these cells take up DNA and express the gene product. Several GAT clinical trials for various diseases including cystic fibrosis, familial hypercholesterolemia, and even cancer and AIDS are currently underway.

Aerosols are used to introduce recombinant viral vector into lungs (so as to bring receptor-mediated endocytosis) for the treatment of cystic fibrosis.

In the treatment of familial hypercholesterolemia, injecting the gene as a sialoglycoprotein complex has increased low-density lipoprotein (LDL) receptor levels.

In cancer gene therapy, a toxin-encoding gene packed in liposomes can be delivered into the cancer cells. In the case of AIDS, appropriate interlukin genes can be delivered to boost the body's immune system.

ii. Targeted gene transfer or gene targeting It is a form of *in vivo* site-directed mutagenesis involving homologous recombination between a targeting vector containing one allele and an endogenous gene represented by a different allele (figure 5.6). Gene targeting can be used either to inactivate a functional endogenous gene or to correct a defective gene. Targeting vectors are of two types:

1. **Integration vectors** (end-in vectors) which are linearised by restriction cleavage within the sequence to be targeted. The cleavage within the homology domain stimulates a single cross over resulting in integration of the entire vector.

2. **Transplacement vectors** (end-out vectors) which have linearisation outside the homology domain or a double cross over event within the homology domain replaces part of the genome with the homologous region of vector.

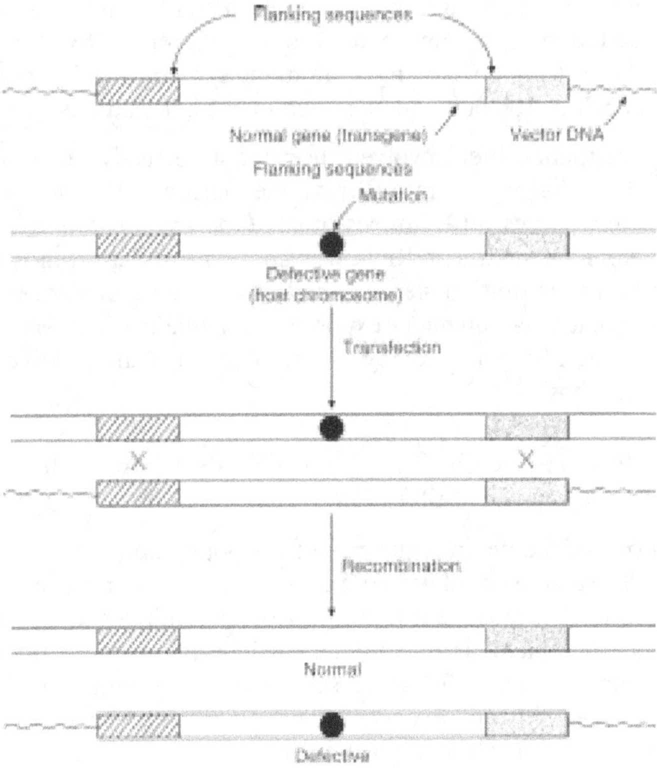

Figure 5.6   Diagrammatic representation of targeted gene transfer through homologous recombination for the replacement of defective gene by a normal function gene

The mammalian viruses specifically retroviruses, adenoviruses and herpes viruses are efficiently used in gene targeting which gives good results because it does not affect the other genes in the genome and the targeted gene is changed in an accurate manner. Yet, there are some technical difficulties like low frequency of homologous recombination, random integration and mutation of transferred gene during transfection. Gene targeting is applied for creation of mutant mice and in gene therapy where a mutant nonfunctional gene is replaced by a normal gene. In near future somatic gene therapy will play an important role in the treatment of several life-threatening diseases.

In April 2003 scientists at Great Ormond Street Hospital in UK, successfully treated an 18 month old child, Rhys Evans, who had a disease called severe combined immunodeficiency (SCID) that prevented him from developing an immune system. SCID occurs in only one birth in 100,000 and is caused by a single mutated gene.

The treatment which involved inserting a correctly functioning version of the faulty gene into Rhys' bone marrow cell is one of the few clinically successful examples of gene therapy anywhere in the world. Before the treatment, Rhys, who comes from Cardiff in Wales, spent most of his time in sterile conditions (inside a bubble) in the hospital so that he would not be exposed to potentially life-threatening infections. After his gene therapy, he was able to run around like any normal little boy.

5.14 IMPROVED CONTRACEPTIVES AND VACCINES TO CONTROL FERTILITY

India is one of the victim countries of the population explosion. In the last 50 years the population of India has increased from 36 crores to 100 crores. If the current trend of growth continues, the natural resources will deplete rapidly and there will be a keen competition to meet the needs. This will lead to several socio-economic problems like shortage of food, accommodation, medical services and employment, and increase in poverty, urban slum, etc. Therefore it is the need of the time to develop contraceptives to have a child by choice and not by chance; so far, in the Indian market, two improved contraceptives are available.

1. Contraceptive pills Saheli developed by Central Drug Research Institute (CDRI), Lucknow that is taken once a week producing no side effects.

2. Contraceptive injection Depo-Provera containing sterile medroxy–progesterone acetate developed by Pharmacia MV/ SA, Belgium. It is administered once in 3 months and does not produce any side effects.

Another approach to develop improved contraceptives involves production of vaccines against Human chorionic gonadotropin (HCG) and antibodies against follicle stimulating hormone (FSH). The vaccine against HCG generates antibodies against HCG. Since HCG is responsible for oocyte development, fertilisation and implantation of embryo into the wall of uterus, the women vaccinated show normal menstrual cycle but pregnancy does not occur due to lowered levels of HCG. Similarly when Anti-FSH antibodies were administered in male monkeys they became sterile without showing any side effects. After successful trials and clinical evolution the products will be released into the market.

5.15 FORENSIC MEDICINE

Medical jurisprudence is the application of the science of medicine in the court of law. In other words forensic medicine is based on the knowledge and techniques of the medical sciences playing a significant role in the resolution of various legal disputes including:

i. Identity of victims of murder, accident, etc.

ii. Confirmation of criminals in the case of rape, murder etc.

iii. Parentage disputes.

DNA fingerprinting or DNA profiling is generally used for identification of criminals and solving parentage disputes. It is taken for granted as a part of modern forensic. Sir Alec Jeffreys of the University of Leicester, UK, discovered the technique in 1980. He found that the gene contained many segments that vary in size and composition from individual to individual and have no apparent function. He called them minisatellites. Jeffreys isolated and inserted these minisatellites into bacteria which produced large amount of DNA segments. They were then purified and labelled with radioactive isotopes to produce the genetic probes that are key to producing genetic fingerprinting.

A brief outline of DNA fingerprinting is as follows:

1. The DNA sample for genetic profiling is obtained from the suspected individual. Usually a blood sample, semen on clothing, vaginal swabs taken from the victim of rape, even the bone marrow isolated from long-buried bones of murder victims can be used as the source of DNA. Researchers can work with a sample as small as one hair root.

2. The DNA obtained from the sample is then digested with suitable restriction enzymes, and the digest is subjected to gel electrophoresis.

3. The denatured DNA is then transferred from gel into a nitrocellulose filter membrane, where it is fixed by backing the filter at 80 °C.

4. The DNA permanently immobilised on nitrocellulose filter is hybridised with the appropriate radioactive probe (usually a single-stranded DNA prepared from minisatellite), the free probe is washed off and the probe hybridised is detected by autoradiography in the form of bands.

Thus the technique of DNA fingerprinting of an individual is essentially a southern blot of his DNA digested with an endonuclease and probed with a radioactive DNA probe. At the end the bands appearing in DNA fragment of suspected criminal/s are compared with those bands obtained from the test DNA recovered from the evidence for crime and believed to originate from the criminals e.g. semen, stains or sperm nuclei in vaginal swabs from rape victims, etc. If the bands match perfectly then it confirms the involvement of the suspected victim in the crime.

Figures 5.7 and 5.8 diagrammatically illustrate the process of DNA fingerprinting and its significant role in identification of criminals and solving parentage dispute.

In case of disputed parentage, the DNA fingerprints of the child, the mother and suspected father are compared. As a rule, the bands present in child's fingerprint must be accounted for by those in the mother's and the father's fingerprints. Figure 5.8 shows the pattern of

DNA fragments of suspected father, mother and 2 daughters whose parentage is disputed. By comparing the bands in DNA fingerprinting of both daughters with their mother and father, it is confirmed that suspected person is father of both daughters.

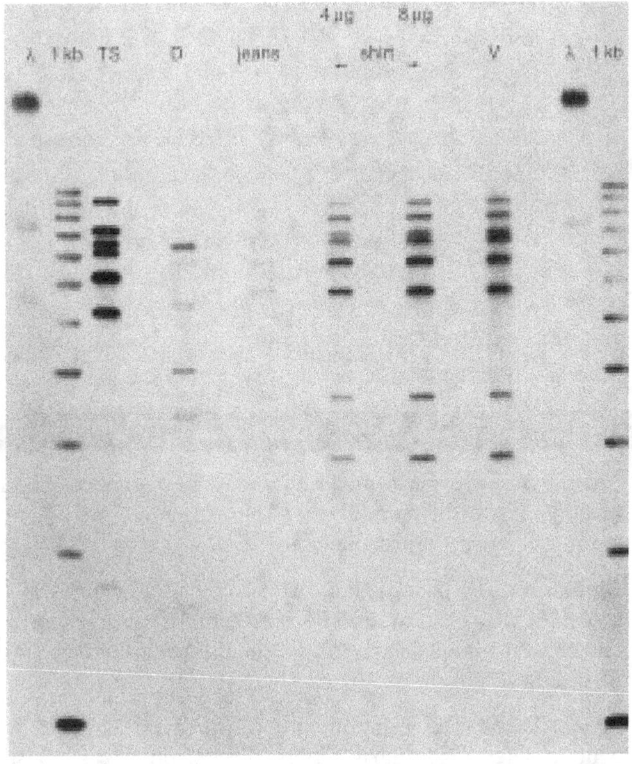

Figure 5.7   DNA fingerprinting. The autoradiograph shown in figure was used in a criminal case in which the defendant was charged with the fatal stabbing of a woman. The bloodstains on the pant and shirt of defendant (D) were compared to the known blood standards from victim (V) and defendant. DNA from bloodstains on the defendant' clothing did match that of victim

DNA fingerprinting has many uses beyond establishing identity or parentage. Consumer protection by DNA profiling tests can be applied to many different products, including wine, food and fabrics for authentication. Combined with archaeology and palaeontology, DNA profiles can also tell about the distant origins of plants and animals. Zoologists use DNA fingerprinting in their efforts to protect endangered species. DNA tests can determine pedigree for seed or livestock breeders.   DNA profiling also involves in other aspects of

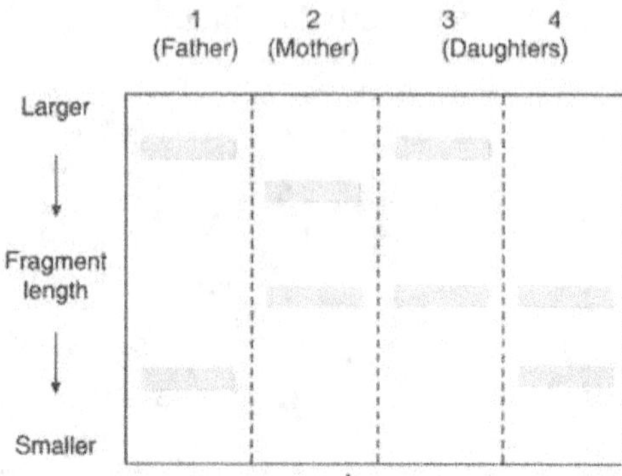

Figure 5.8   Role of DNA fingerprinting in solving parentage dispute of daughters

---

Key Concepts

- Interferons are glycoproteins providing protection to healthy cells form virus attack. They stimulate the body's natural defence against the cancer cell.

- Biotechnology-based diagnostic tools, including DNA probe, ELISA and immuno-PCR are simple, more sensitive, and rapid methods that provide authentic results in diagnosis of diseases.

- Amniocentesis is a technique used for prenatal diagnosis of genetic defects. Similarly biopsy of chorionic villi is also helpful for study of chromosomal and biochemical disorders.

- Enzyme or gene replacement therapy is used to combat the various genetic diseases like SCID, Gaucher's disease, familial hypercholesterolemia, cystic fibrosis etc.

- Gene targeting can be employed either to inactivate a functional endogenous gene or to correct the defective gene using targeting vectors like integration or transplacement vectors.

- Forensic medicine exploits DNA fingerprinting or DNA profiling discovered by Jeffreys for identification of victims, confirmation of criminals and parentage dispute.

medicine such as in genetic counselling, tracing the frequency of donor cells in bone marrow transplants, tissue culture cell line identification, etc. It is also applied in animal husbandry for the proof of parentage, pouching, identification in cases of theft or loss, detection of trait markers, etc. It is also applicable to cats, dogs and birds.

In India the medico-legal cases related to Tandoor case of Naina Sahani's murder and the parentage dispute of the child belonging to Amarmani Tripathi and the murder of Madhumita which attracted much of the electronic media were solved on the basis of DNA fingerprinting. Even the FBI of USA has confirmed Saddam Hussein in Iraq by doing DNA profiling by taking his blood sample compared with DNA samples obtained from his two dead sons. Thus DNA profiling is a novel technique playing an indispensable role in modern medicine by providing fast and reliable results than any conventional identification technique.

Review Questions

1. Explain the contribution of biotechnology to diagnosis, prevention and cure of human diseases.

2. What are antibodies? How are they classified? Enlist the role of monoclonal antibodies.

3. Define vaccine. Explain the role of vaccines in the cure of hepatitis B and AIDS.

4. What are antibiotics? Discuss various antibiotics with their potential role in disease control.

5. Define hormone. Briefly explain the process of production of insulin and human growth hormones with their role in human health care.

6. What is a cancerous cell? Explain the role of different interferons in the control of cancer.

7. What is gene therapy? What are the types of gene therapy? Explain their role in the treatment of genetic diseases.

8. Define forensic medicine. Explain the process and role of DNA profiling in forensic medicine.

9. Write short notes on:

    a. Improved contraceptives   i. Gene targeting

    b. Amniocentesis   j. Non-radioactive labelled probes

    c. Autoantibodies

    d. DNA probes   k. Fertility control by vaccine

    e. ELISA   l. Therapeutic agents from genetically engineered organisms

    f. Immuno PCR

    g. IgG   m. DNA fingerprinting

    h. Penicillin G

6

ENZYME TECHNOLOGY

6.1 INTRODUCTION

Enzymes are complex protein molecules present in living cells, where they act as biological catalysts and bring about chemical changes in substances. Virtually all the chemical reactions occurring in microorganisms, plants and animals proceed at a measurable rate as a direct consequence of enzymic catalysis. Without enzymes there can be no life. Enzymes can change plants and animals in a precise and often remarkable dramatic fashion. In the hands of expert biotechnologists enzymes become the tailor's scissors and the surgeon's scalpel. Although enzymes are formed only in living cells, they can be separated from the cells and can continue to function *in vitro*. This unique ability of the enzymes to perform their specific role in an isolated system formed the basis of an ever-increasing use of enzymes in industrial processes (medicine, research and rDNA technology) that are collectively known as enzyme technology.

Enzyme technology involves the production, isolation, purification, use in soluble form and finally immobilisation of enzymes and also their role in various industries including dairy, baking, brewing, textile, food production, leather, pharmaceutical and medicine. In addition, enzyme technology has immense importance in development of biosensors that have tremendous role to play in world economy.

6.2 HISTORICAL BACKGROUND

W. Kuhne coined the word 'enzyme' in 1878 from the Greek term meaning yeast. Earlier enzymes were referred to as ferments because their action was similar to yeast fermentation. Some of the chronological events in the field of enzyme study are listed in table 6.1, indicating that during the succeeding years enzymology has developed apace.

Table 6.1   Chronology of enzyme study

1833	Payen and Persoz achieved alcohol precipitation of thermolabile 'diastase' from malt.
1835	Berzelium put forth the concept of catalysis.
1850	Wilhelmy studied quantitative evaluation of rates of sucrose inversion.

Table 6.1   *Contd...*

1878	Kuhne investigated trypsin-catalysed reaction, and introduction of the word 'Enzyme'
1890s	Fisher suggested the 'Lock and Key' model for enzyme action.
1898	Duclaux suggested nomenclature for the enzymes with substrate plus suffix 'ase'
1906	Harden and Young studied coenzymes (NAD)
1913	Michaelis and Menton proposed mathematical model for kinetic theory of enzyme action.
1926	Summer isolated enzyme urease in crystalline form.
1937–39	Cori and Cori studied muscle phosphorylase.
1940	Beadle and Tatum put forth 'one gene –one enzyme' hypothesis.
1948	Pauling. Transition state theory of enzyme action.
1951	Pauling and Corey. Secondary structure of enzyme.
1953	Koshland. Induced fit hypothesis
1952	Sanger. Amino acid sequence of protein hormone, insulin.
1956	Sutherland. Cyclic AMP- second messenger.
1961	Jacob, Monod and Changeux. Allosterism.
1986	Cech. Ribozyme-RNA with catalytic activity.
1994	Cech and Uhelnbeck. Hammer headed ribozyme to fight viral diseases.
1997	Tang and Breaker. Rational design of allosteric ribozymes.
1999	Robertson and Ellington. *In vitro* selection of an allosteric ribozyme.
2000	Bergman *et al.* Kinetic framework for ligation by an efficient RNA ligase ribozyme.
2001	Seetharaman *et al.* Ribozyme in ELISA –like assay.
2002	Iyo *et al.* Allosterically controllable maxizymes for molecular gene therapy.
2003	Vaish *et al.* Half-ribozyme activated by hepatitis C virus (HCV) sequences.

6.3 TERMINOLOGY USED IN ENZYMOLOGY

Endozymes   The enzymes acting within the cell in which they are synthesised.

Exoenzymes   The enzymes acting outside the cell in which they are synthesised.

Purely proteinaceous enzymes   These are made up of only proteins e.g. amylase, proteases.

Conjugate enzymes   These enzymes are made up of protein molecules to which a non-protein group (prosthetic group) is attached e.g. oxidising enzyme.

Substrate   The substance upon which enzyme acts.

Prosthetic group   It is non-protein part of enzyme found attached to protein part of the enzyme. e.g. haem, biotin, pyridoxal phosphate, etc.

Apoenzyme   It is an enzyme whose prosthetic group is removed.

Coenzyme   These are low molecular weight organic molecules that combine with enzyme. e.g. vitamins, NAD, FAD, DPN, Coenzyme A, etc.

Cofactor: These are integral parts of the enzymes in the form of metal ions like Mg^{++}, Fe^{++}, Cu^{++}, Zn^{++}, etc.

Enzyme activation   An expression of the ability of a given enzyme preparation to catalyse a specific reaction. It may be defined in terms of number of moles of substrate converted, or the number of moles of product produced, in unit time per unit weight of protein. (e.g. micromoles per milligram protein per minute)

Enzyme assay   a method for determining the activity of an enzyme sample.

Enzyme extraction   The removal of enzymes from contaminating materials in order to increase their specific activity.

Enzyme fermentation   A process in which a microorganism (bacterium, fungus, or yeast) is grown for production of an industrial enzyme on a large scale.

Enzyme immobilisation   The conversion of a soluble enzyme to a bound or insoluble form.

Enzyme induction The synthesis of enzymes in response to an inducing agent by stimulating expression of the genes encoding the protein with a specific enzyme activity.

Enzyme inhibition A mechanism by which an enzyme is inactivated by a chemical agent.

Enzyme repression It occurs in many anabolic systems, in which the production of the enzymes of a synthetic pathway is selectively inhibited by the end product of that metabolic chain.

Isoenzymes These are multiple forms of an enzyme that differ by minor variations in amino acid composition and sometimes in regulation. e.g. LDH.

Extremozymes There are enzymes, which function optimally under extreme conditions of temperature, pH, etc. There are about 30 enzymes that have been isolated from hyperthermophile (optimal growth at or above 100 °C) sulphur-reducing bacteria such enzymes are also known as high temperature enzymes and they have significant role in Polymerase Chain Reaction work.

6.4 ENZYME CLASSIFICATION AND NOMENCLATURE

Enzymes are the largest and most specialised class of protein molecules that catalyse a reaction in which a substrate (S) is converted to product (P) through the formation of an intermediate enzyme–substrate (ES) complex. On the basis of the type of reactions that they catalyse, enzymes are classified into six main classes and named by the Commission on Biochemical Nomenclature of the International Union of Biochemistry as follows:

1. Oxidoreductases These are enzymes catalyzing oxidation and reduction reactions by the transfer of hydrogen and/or oxygen. e.g. glucose oxidase

$$AH_2 + B \rightleftharpoons A + BH_2 \text{ (Reduction)}$$
$$AH_2 + O_2 \rightleftharpoons A + BH_2 \text{ (Oxidation)}$$

2. Transferases These enzymes catalyse the transfer of certain groups between two molecules. e.g. aspartate aminotransferase.

$$A - X + B \rightleftharpoons A + B - X$$

3. Hydrolases These are enzymes catalysing hydrolytic reactions. This class includes amylases, proteases, lipases etc. eg. renin, lysozyme.

$$A - B + HOH \rightleftharpoons AH + BOH$$

4. Lyases These enzymes are involved in elimination reactions resulting in the removal of a group of atoms from substrate molecule to leave a double bond. It includes aldolases, decarboxylases, and dehydratases, e.g. fumarate hydratase.

$$A = B + X - Y \rightleftharpoons AX - BY$$

5. Isomerases These enzymes catalyze structural rearrangements within a molecule. Their nomenclature is based on the type of isomerism. Thus these enzymes are identified as racemases, epimerases, isomerases, mutases, e.g. xylose isomerase.

$$A \rightleftharpoons A'$$

6. Ligases or synthetases These are the enzymes which catalyse the covalent linkage of the molecules utilising the energy obtained from hydrolysis of an energy-rich compound like ATP, GTP e.g. glutathione synthetase.

$$A + B + ATP + H_2O \rightleftharpoons A - B + ADP + Pi$$

6.5 BIOLOGICAL SOURCES FOR ENZYME PRODUCTION

Various microorganisms like bacteria, fungi and yeast produce different kinds of enzymes; even the animal tissues and plants are used as sources of enzymes (table 6.2). Initially plant and animal enzymes were preferred over microbial enzymes mainly because of safety concern and fear of contamination by microorganisms, toxins, etc. But as the demands for enzymes augmented, the supply of enzymes derived from plant and animal tissues could not keep the pace. Later on the importance of microbial enzymes increased because of the advantage of (a) large-scale production by fermentation, (b) ease in isolation, and (c) involvement of recombinant DNA technology so that the quantity and quality of enzymes can be modified. For example, the enzymes rennet (also called chymosin or aspartic proteinase) is widely used for cheese production; in earlier days, rennet was obtained from animal tissue like calf stomach. Since it is difficult to maintain a continued supply of rennet derived from animal tissues, nowadays it is produced on large scale from the fungal species, *Mucor meihei.*

Table 6.2   A list of biological sources for production of industrially important enzymes with their applications

Source	Tissue/Producer Microorganism	Enzyme	Application
Animal tissue	Liver	Catalase	Food
	Pancreas	Lipase	Food
	Pancreas	Trypsin	Leather and medicine
	Stomach	Rennet	Dairy
Plant	Malted barley	α-amylase	Food, textile and brewing
	Malted barley	β-amylase	Brewing
	Malted barley	β-Glucanase	Brewing
	Fig latex	Ficin	Food
	Papaya latex	Papain	Meat
Bacteria	*Bacillus*	α-amylase	Starch, textile and biological detergent
	Bacillus	β-amylase	Starch
	Bacillus and *Streptomyces*	Glucose isomerase	Fructose syrup
	Bacillus	Penicillin amidase	Pharmaceutical
	Bacillus	Protease	Detergent and Brewing
Fungi	*Aspergillus*	α-amylase	Baking
	Aspergillus and *Rhizopus*	Glucoamylase	Starch
	Mucor meihei	Rennet	Dairy
	Aspergillus	Pectinase	Fruit juices
	Aspergillus	Protease	Brewing and baking
	Aspergillus	Cellulase	Fruit juices
	Rhizopus	Lipase	Detergents
	Aspergillus	Lactase	Dairy
Yeast	*Saccharomyces*	Invertase	Confectionery
	Saccharomyces	Raffinase	Food

6.5.1 Advantages of Microbial Enzymes

Microbial enzymes are commercially better than plant or animal enzymes due to two types of advantages—economic and technical.

Economic advantages

The sheer quantity of enzyme that can be produced by microbes in a short time and in a small area is much greater than that of animal- or plant-derived enzymes. For example, a 1000 dm^3 fermenter of *B. subtilis* can produce up to 20 kg of enzyme in 12 hours. The average quantity of rennet extracted from one calf stomach is 10 g and it takes several months to produce one calf. Thus one gains from the economics of scale. The second advantage is the ease of extraction. A large proportion of industrial enzymes, hydrolases from microbial sources for example, are produced extracellularly into growth medium. Thus there are no troublesome extraction problems. Even in the case of intracellular microbial enzymes, extraction usually involves fewer steps than for plant and animal enzymes. For example the first step in the production of enzymes from plant or animal enzymes involves the harvesting of the organism and its transportation to the processing plant. Microorganisms are usually produced on the spot (with few exceptions) thus not only saving on the cost of harvesting and transport but also allowing the integration of the process of production and extraction of enzyme. Equally important is the consideration that animal or plant enzymes, unlike that of the microorganisms, are usually specifically located in a particular tissue or organ and that this portion of the organism must first be separated from the rest and remains disposed of. Further, the predictability of enzyme yield and the absence of seasonal variation are some more economic advantages in microbial enzymes. Plant or animal enzyme sources are subject to wide variations of yield and may be available only at certain times of the year.

Technical advantages

There are four technical advantages in employing microbes as enzyme producers rather than plants or animals. The first is the enormous variation of biochemical pathways and hence enzymes, which exist not only across the world of microorganisms but also within individual species. Thus one microorganism is theoretically capable of producing a wide variety to different enzymes. The second advantage of microorganisms is the range of environments in which they grow. Enzymes can remain stable under extreme environmental conditions and are more likely to be found in microorganisms which are adapted to grow in extreme environments. The best example of this is the thermophilic organisms that often yield enzymes that are themselves stable at high temperatures. The third advantage is the genetic flexibility of microorganisms that results in the relative ease with which microorganisms may be manipulated to increase the yield of enzyme. The traditional techniques for the yield improvement, strain or mutant selection, induction, depression, or alteration of the growth medium are still preferred to gene transfer. This is a low-cost technology, and a rapid, proven method that does not depend on knowledge of genetics or biochemistry of a donor organism. Last but not the least, the short generation time of microorganisms is a great advantage. The average bacteria reproduce in a matter of minutes, whereas plants may take weeks and animals may take months to double their size. Above all, with the advent of recombinant DNA technology, when an enzyme has been identified as a good candidate for industrial use, the relevant gene can be cloned into a more suitable production host microorganism and industrial fermentation can be carried out. Thus it becomes possible to produce very high quality and purity of industrial enzymes using microorganisms rather than using plants or animals as enzyme producers.

6.6 RECOMBINANT DNA TECHNOLOGY FOR ENZYME PRODUCTION

Genetic engineering for industrial production of enzymes is possible by transfer of the gene/s encoding useful enzymes into a suitable host microorganism (figure 6.1). Introducing more copies of gene into the concerned organism may increase the level of production of an enzyme. A recent example of this technology is the detergent enzyme lipolase produced by Novo Nordisk A/S, which has improved removal of fat stains in fabrics. The enzyme was first identified in the fungus *Humicola languinosa*. The DNA fragment for the enzyme was cloned into production fungus *Aspergillus oryzae* and commercial levels of enzyme production achieved. The enzyme is very stable at a variety of temperatures and pH conditions relevant to washing. In addition, lipolase is remarkably resistant to proteolytic activity of the commonly used detergent proteases.

The modification in the catalytic activity for the usefulness of the existing enzyme or production of new enzyme activity by making suitable changes in its amino acid sequence is known as **enzyme engineering**. Since all enzymes are proteins, enzyme engineering is an integral part of protein engineering. The main objective of this approach is to modify various properties of the enzyme, so that enzyme become more useful.

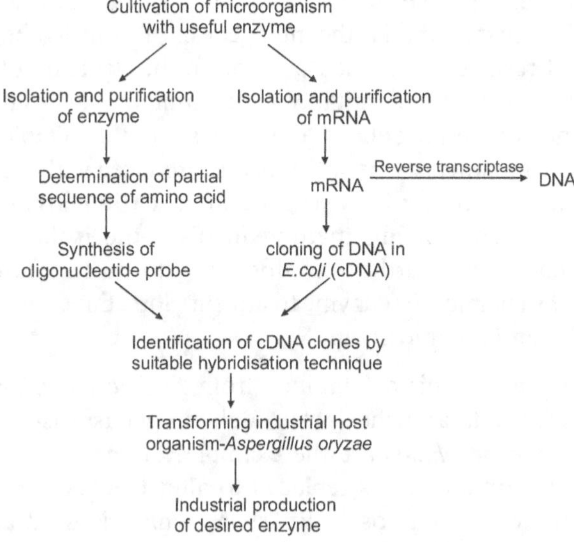

Figure 6.1   Genetic engineering for industrial enzyme production

The main objective of enzyme engineering is to produce an enzyme which is more useful in industrial and other applications by modifying them in order to:

- improve the activity of enzyme
- enhance the stability
- alter optimal pH and temperature
- increase thermostability
- alter the specificity of an enzyme so that it catalyses the conversion of different substrate
- improve the efficiency of a process

The alteration in the properties of an enzyme is always reflected in its primary structure i.e. the amino acid sequence. **Protein engineering** is rightly called molecular surgery as it introduces the amino acid changes in certain critical regions of the protein. The protein engineering of enzyme in achieved by developing a three-dimensional graphical model of purified enzyme with the help of x-ray crystallography, nuclear magnetic resonance (NMR), etc. Using this data, a molecular model is prepared for determining a possible change in the sequence of enzyme. This can be achieved by two methods. The first method is the mutagenesis of cloned-gene product. Amino acid residues at defined position in the structure of enzyme can be replaced by other suitably coded amino acid residues. The altered gene is then introduced and expressed in a suitable host i.e. *E.coli* and the mutant enzyme subsequently produced with the requisite changes in position. This approach is called **site-directed mutagenesis.** The second method involves the isolation of the natural enzyme and further modifications to its structure can be carried out by chemical or enzyme treatment. Sometimes this approach is called **chemical mutation**.

The enzyme subtilisin (obtained from *Bacillus amyloliquefaciens*) used in detergents and the lactate dehydrogenase (isolated from *Bacillus stearothermophilus*) are the examples of enzyme engineering by which the amino acid is replaced to alter their performance. In addition the enzyme phospholipase A, which is used as a food emulsifier, was modified structurally to resist higher concentrations of acid. Genetic engineering and protein engineering will have a significant role to play in the enzyme industry, since they ensure better product economy, production of enzymes from rare microorganisms, faster development programmes and above all, such enzymes are ecofriendly and have low allergenic potential.

6.7 SYNZYMES OR ARTIFICIAL ENZYMES

These are synthetic polymers that may be protein or non-protein molecules having catalytic activities. Synzyme has also two functional sites, one substrate binding site and a catalytically effective site. The artificial enzymes also follow kinetic properties just as natural enzymes do. Examples of Synzymes are –

1. Derivatised myoglobin The conjugate protein, myoglobin, is present in muscles and is an oxygen carrier. When the group $[RU(NH_3)_5]^{3+}$ is attached to three surface histidine residues of this carrier protein molecule, it begins to work as an oxidase (originally it was oxygen carrier), it oxidises ascorbic acid and reduces molecular oxygen. Thus, this derivative of myoglobin functions as effectively as natural ascorbate oxidase.

2. Cyclodextrin is a nonprotein synzyme and is a naturally occurring cyclic oligosaccharide molecule having α, β, γ, δ, and ε forms. It can be attached to several catalytically active groups. For example, when pyridoxal coenzyme is attached to a C-6 hydroxyl group of β-cyclodextrin it acts as a transaminase, which is not so active as natural transaminase, specific for L-amino acids. Similarly, alkylated polyethylene amine is a non-protein synzyme.

6.8 RIBOZYME

These are the modified RNA molecules that have the capability of catalysing certain chemical reactions. Ribozymes function as true catalysts enhancing the rate of chemical reactions without any net change to themselves. The catalytic ability of ribozymes arises due to their three-dimensional structures (figure 6.2) which enables them to generate the substrate-specific binding site, similar to enzymes.

Ribozymes possess two properties that make them potentially important agents in fighting against certain chronic diseases in humans, animals and plants.

1. They contain stretches of nucleotides that allow them to base-pair with a complementary RNA and

2. They have a catalytic site (as mentioned above) that is able to cleave the backbone of the complementary RNA. With these properties, ribozymes are ideal subjects for manipulation by genetic engineers because they can be custom-designed to recognise and destroy any specific RNA target. The custom-designed ribozymes are used in the fight against viral diseases. Recent experiments carried out by Wongstaal and Hempel of University of California and Northern Illinois University respectively, have demonstrated the feasibility of using ribozymes in the fight against diseases caused by RNA viruses.

In their experiments, they isolated T- lymphocytes from patients with AIDS and transfected the cells with a genetically engineered DNA sequence that encodes a ribozyme capable of recognising and cleaving HIV mRNA. It is hoped that transfected cells will prove resistant to the virus and preserve the patient's crippled immune system. The ultimate goal in this type of gene therapy is to introduce the gene for an RNA destroying ribozyme into all cells of the body that are infected by the virus, rather than simply those that circulate in the bloodstream.

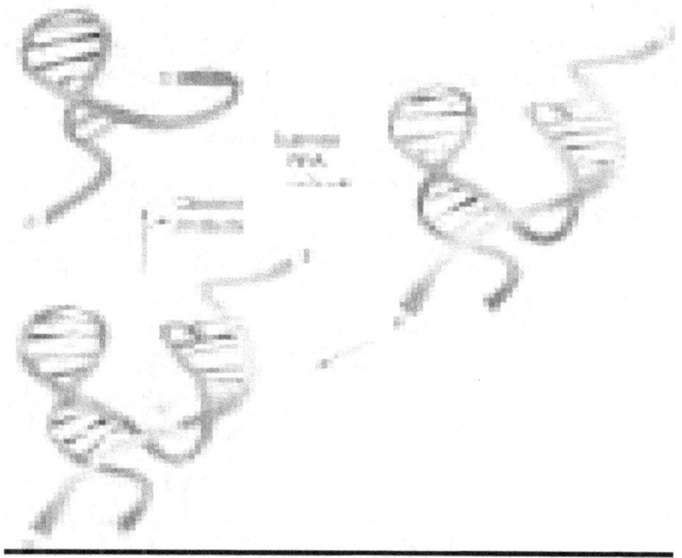

Figure 6.2   Hammer-headed ribozyme showing catalytic activity

Ribozymes may also prove useful in the fight against certain types of human cancers. The mutation in an oncogene that encodes a protein involved in regulative cell division results in human cancer. Since the malignancy to these cells due to synthesis of an RNA that is different from that produced by normal cell, it should be feasible to reverse the malignant state of the cell by destroying the altered RNA with the help of ribozyme. To be effective as a treatment for cancer, the gene encoding the ribozyme would have to be introduced into all the cells of the tumour. This requires the development of a vector that is capable of delivering the gene to the body's internal tissues.

Ribozymes are also used to engineer plants for virus-resistance. Transgenic tobacco plants expressing ribozymes against Tobacco Mosaic Virus showed some resistance to TMV infection.

6.9 IMMOBILISED ENZYMES

Immobilisation of an enzyme has been defined as the process by which the movement of an enzyme in space is completely or severely restricted, usually resulting in a water-insoluble form of the enzyme. The process can be extended to cover cells and organelles. The immobilisation can be achieved by fixing the enzyme and other molecules to or within some suitable material. The carrier matrices are inert polymers or inorganic materials used for immobilisation of enzymes. An ideal carrier matrix should be economical, inert, stable and regenerable. In addition, it should increase the enzyme specificity and reduce the production inhibition.

6.9.1 Methods of Immobilisation

Depending upon the nature of physical relationship of an enzyme to the polymer matrix, the methods of immobilisation can be divided into various groups. Thus the enzyme may be covalently bonded to the polymer, physically adsorbed onto the polymer, cross linked with itself, entrapped inside a carrier matrix or encapsulated in a 'polymer bag' (figure 6.3). Moreover any combination of these methods can be applied to immobilise the biological catalyst. In all the above-mentioned methods, the polymer, polymer matrix or carrier matrix employed for the effective immobilisation of the enzyme is in the form of material such as cellulose hydrogels, nylon, glass, polyacrylamide beads or even iron filings. Despite the diversity of the methods of immobilisation, the following four methods are generally used for the same: (a) covalent bonding, (b) cross linking, (c) adsorption, (d) entrapment, and (e) encapsulation.

Covalent bonding In this method the catalysts are attached to the polymer matrix by the formation of covalent bonds. This can be achieved by two ways, the first by activating the polymer with a reactive group, the second by the use of a bifunctional reagent to link an enzyme with the polymer. Hydroxyl, amino groups or sulphydryl groups of an enzyme are involved in covalent bonding with polymer molecule. Since the strength of bonding is very strong it does not lead to loss of enzyme during use.

The large number of polymer matrices used in this process are inorganic carriers such as ceramics, glass, iron, zirconium and titanium, the natural polymer such as sepharose and cellulose, and the synthetic polymers such as nylon, polyacrylamide and other vinyl polymers and copolymers.

The disadvantage of this technique is that the enzyme may often be inactivated due to conformational changes in the enzyme's active site, which can be overcome by performing immobilisation in the presence of the enzyme's substrate or a competitive inhibitor, or with proteases. Hydrogels are the commonest polymer matrices that can be activated by the treatment with cyanogen bromide to which biological catalysts bind to form immobilised enzymes. The use of bifunctional reagents such as gluteraldehyde that exist in the equilibrium mixture of monomers and oligomers provides an alternative strategy for covalent enzyme immobilisation. In addition to frequent inactivation of an enzyme, covalent bonding has disadvantages in the form of use of toxic reagents, complicated preparative procedures, and high costs involved.

Cross-linking   The covalent bonding technique can be extended to the cross-linking of the enzyme with the bifunctional reagent. In this approach, bifunctional reagent molecules such as gluteraldehyde bind to two enzyme molecules and a network of enzyme molecules linked together is produced. The disadvantages of this method include loss of enzyme activity during preparation, ineffectiveness for macromolecular substrates, and difficulty in regeneration of carrier molecule. However, the technique is cheap, simple, and widely used in commercial preparations of immobilised enzymes such as glucose isomerase that is used in food industry.

Adsorption   Perhaps this is the most widely used technique for immobilisation of enzymes. It is the simplest and cheapest method where the enzyme or cell adsorbs to a polymer material due to combination of hydrophobic interactions and the formation of several salt links per enzyme molecule. The method has additional advantages in that there is no modification of enzyme and regeneration of carrier molecule is possible. However, the technique has its own drawbacks— changes in ionic strength may cause desorption and the enzyme is subjected to microbial or proteolytic enzyme attack. The process requires careful selection of matrix. The most commonly used matrices are, ion- exchange matrices, porous carbon, clays, hydrous metal oxides, glasses and polymeric aromatic resins. The technique involves most

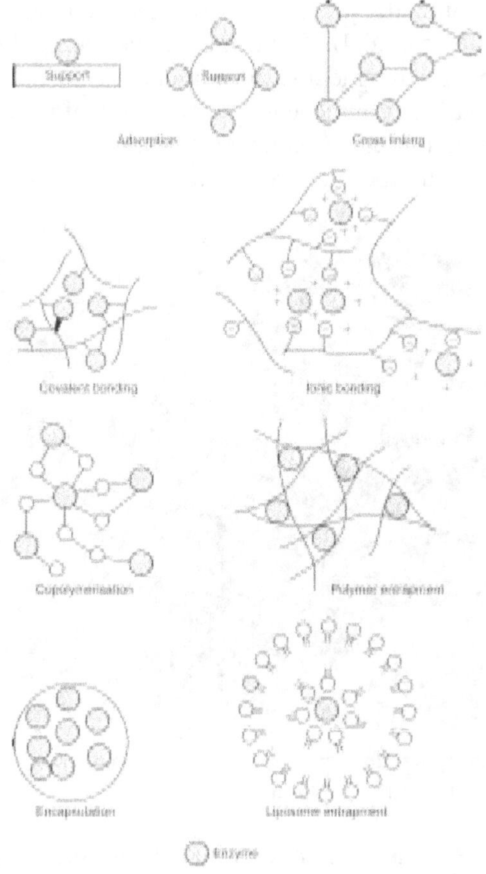

Figure 6.3　Different methods of immobilisation of enzyme

commonly employed adsorbant in the form of ion exchange cellulose (such as DEAE cellulose), to which an enzyme is mixed under appropriate conditions of pH and ionic strength. After incubating this mixture for a sufficient period of time, the carrier is washed to eliminate unadsorbed enzyme molecules, and the immobilised enzyme is available for use.

Entrapment　In this method, catalysts are held or entrapped within appropriate gels or fibres. The biocatalyst is dissolved in a solution of polymer precursors and polymerisation initiated. The widely used polymers are polyacrylamide gels, cellulose acetate, agar, gelatin or alginate. Entrapment of an enzyme or microbial, animal and plant cells within calcium alginate is the most widely employed method of immobilisation (figure 6.4). The technique has some disadvantages. Continuous loss of enzyme due to variability of the pore size in gels, ineffectiveness in macromolecule substrates and diffusion of substrate to the enzyme and of the product away from the enzyme, make the preparation difficult and often result in enzyme inactivation.

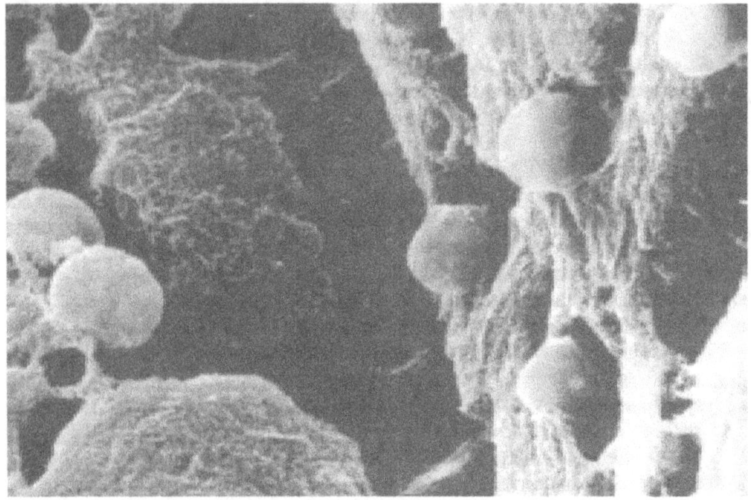

Figure 6.4   Scanning electron micrograph of algal cells immobilised
            using the matrix of calcium alginate

Encapsulation   This is a method in which the biocatalyst, usually in an aqueous solution, may be enclosed in a semipermeable membrane capsule. Encapsulation allows free movement in either directions to the substrates and products but prevents the escape of biocatalyst from the capsule. The technique is simple and cheap but an important condition is that enzyme must be stable in solution. The biodegradable material in the form of polylactic acid or phospholipid liposomes can be used for encapsulation of enzyme. The major disadvantages of the technique is that the enzyme may lose its stability and may undergo denaturation. The molecular weight of the substrate is also influencing factor that minimises the applicability of this technique.

In spite of the disadvantages mentioned in various techniques, the immobilised enzymes have a number of advantages over the free biocatalysts.

1. Immobilised enzymes are normally more stable than their soluble counterparts.

2. They can be reused in the purified, semi-purified or whole-cell form.

3. Catalytic properties of immobilised enzymes can be altered favourably and this permits the enzyme to function under broader or more rigorous reaction conditions; increase in thermostability is one such favourable change in immobilised enzymes. e.g. An immobilised glucose isomerase can be used continuously for over 1000 hours at temperatures of between 60 and 65 °C; while its normal form denatures at 45 °C.

4. Since the product can be readily separated from the enzyme, it effectively saves the cost of downstream processing of the product.

5. Immobilisation of an enzyme makes recovery easy and may also reduce effluent-disposal problems.

6. It allows more accurate control of catalytic processes.

7. It provides an ideal system for continuous operation.

8. It offers considerable potential for industrial and medical use.

6.9.2 Effects of Immobilisation on Enzymes

During the process of immobilisation of an enzyme, the objective is to achieve as high an active biocatalyst loading per unit volume of immobilisation support (polymer matrix) as possible. The different parameters normally measured are the volume, enzyme activity, and protein content or viable cell count of the enzyme or cell used, the weight, particle size distribution, porosity and the chemical and physical properties of polymer used, the activity and protein concentration of any biocatalyst remaining after completion of immobilisation, operational stability, and productivity and resistance to microbial contaminations.

Once the immobilisation of an enzyme is completed, the kinetic behaviour of an immobilised enzyme differs significantly from that of its free counterpart. Different enzymes show different response to the same immobilisation technique. Therefore it is essential to employ a suitable immobilisation technique for the given enzyme.

Immobilisation techniques may induce or suppress the stability of an enzyme; it may put a strain in the enzyme molecule. The ultimate effect of this strain is that the enzyme becomes vulnerable to denaturation by higher temperatures, pH, etc. The other effects of immobilisation could include total or partial inactivation caused by gross conformational change or reaction of some essential group at the enzyme's active site, more subtle induced conformational change causing destabilisation, alteration of allosteric effects or kinetic characteristic stabilisation. Limitation of the free diffusion of solute molecules by the physical presence of polymer matrix will cause the decrease in substrate molecules and concentration of the product molecules around the enzyme. The pH 'profile' of an enzyme (i.e. a graphical plot of the rate of enzyme activity versus pH) may be displaced, distorted, broadened or narrowed as a consequence of immobilisation. An electrostatic charge on the polymer matrix may have a direct effect on the stability of ionising groups at the enzyme's active site thus raising or lowering their pKa value. In spite of these variations in enzyme kinetics, if an enzyme molecule is subjected to multipoint binding, it may not create any strain on it, which ultimately leads to substantial stabilisation.

6.9.3 Uses of Soluble Enzymes

Enzymes have an enormous range of applications in industry, medicine, research, etc. The various uses of enzymes in solution are briefly explained as follows:

1. Biological detergents represent the largest industrial application of enzymes. The chief enzymes utilised to manufacture detergents, soaps and oxidants are as described below:

   i. Proteases are used to digest proteins present in bloodstains, milk, grass, dirt, etc. These enzymes are produced by *Bacillus* species. The problem of allergic response of the workers and users which was encountered before now is overcome by encapsulation techniques.

   ii. α-amylase enzymes, produced by *Bacillus* species are used to digest starch present in association with stains and/or dirt.

iii. Cellulases produced by fungi are used for washing cotton fabrics.

2. The baking industry uses the following enzymes

i. Fungal α-amylase enzymes catalyse breakdown of starch in the flour to sugar, which can be used by the yeast. It is used in the production of white bread, buns, rolls, etc.

ii. Proteinase enzymes used in biscuit manufacture to lower the protein level of the flour.

3. Brewing industry uses the following enzymes

i. α- and β-amylases and β-glucanase produced from barley; these enzymes are utilised to degrade starch to produce simple sugars used by the yeasts to enhance alcohol production.

ii. Amyloglucosidase and proteinases split polysaccharides and proteins in the malt and use in production of low-calorie beer and remove cloudiness during storage of beers.

4. Dairy industry utilises the following enzymes

i. Renin obtained from the stomach of young ruminant animals like calves, lambs or kids and from fungus *Mucor meihei* is used in the manufacture of cheese.

ii. *Aspergillus* derived lactase used to breakdown lactose to glucose and galactose.

iii. Lipases produced by *Rhizopus* are employed for cheese ripening and in icecream manufacture.

5. Textile industry uses the enzymes such as

i. α- and β-amylases derived from *Bacillus* species are widely utilised to remove starch which is used as desizing and binding agents on threads of certain fabrics to prevent damage during weaving.

ii. Glucoamylases obtained from *Aspergillus* and *Rhizopus* are generally preferred for desizing, since they are able to withstand working temperatures upto $100–110\,^{0}C$.

6. Confectionary, softdrink and food industries is based on the involvement of following enzymes-

i. Heat stable fungal proteases used for hydrolysis of gluten wheat, make dough suitable for biscuit, pie and pastry making.

ii. Amylases (α- and β-), glucoamylases, invertase and glucose isomerases obtained from microbes act on starch, sucrose and D-glucose to produce glucose, maltose and high fructose syrups.

iii. Papain derived from papaya latex is utilised for tenderisation of meat.

iv. Glucose oxidase derived from *A. niger* and *Penicillium,* catalyses the formation of gluconic acid from β-D glucose. The enzyme is used for removal of glucose or oxygen from food in order to increase its storability. In the process hydrogen peroxide (H_2O_2) is produced which effectively kills bacteria.

v. Catalases derived from animal tissues are used to degrade H_2O_2 into water and oxygen. It is usually applied in combination with glucose oxidase to remove glucose and/or O_2 from foods, drinks, etc.

vi. Pectinases and cellulases obtained from *Aspergillus* are used to reduce viscosity and increase juice yield with enhanced flavour.

7. Leather Industry

Traditionally enzymes found in dog and pigeon faeces were used to treat leather to make it pliable by removing certain protein components the process is called bating; strong bating required to achieve a soft, pliable leather, slight bating for the soles of shoes. Bating is also essential for production of soft leather clothing. Nowadays trypsin an enzyme obtained from slaughterhouses and from microorganisms is utilised for removing the hair from hides and skins to increase suppleness and softness in appearance.

8. Pharmaceuticals and medical industries

The trypsin, obtained as mentioned above, is used in debridement of wounds, dissolving blood clots. The pancreatic trypsin is utilised for digestive aid formulations and treatment of inflammations. Many enzymes are used in clinical chemistry as diagnostic tools some of which are employed in treatment of diseases.

i. α-Amylases, lipases, proteases are used in treatment of digestive disorders.

ii. Asparaginase, glutaminase are utilised for treatment of cancer, specifically leukemia.

iii. Ribonuclease is used as antiviral agent.

iv. Rhodonase is used for cyanide poisoning.

v. β-lactamase is used in treatment of penicillin allergy.

vi. Streptokinase and urokinase for treatment of the disorders of blood circulation.

6.9.4 Uses of Immobilised Enzymes

Several industrial processes are based on the use of immobilised enzymes, as it is advantageous than enzymes in solution. Some of the common industrial applications of immobilised enzymes are as follows.

1.Food industry utilises immobilised glucose isomerase from *Actinoplanes missouriensis*, *Bacillus coagulans* and *Streptomyces* sp. This enzyme is used to convert glucose syrup into high fructose syrup, which is exploited for preparation of soft drinks, since fructose is sweeter than glucose. Immobilised aminoacylase is another enzyme employed for resolution of D, L- amino acids to L-form.

2. In dairy industry immobilised lactase is used to hydrolyse lactose to glucose and galactose. Since many people are sensitive to lactose, lactase is used to remove the lactose from milk and whey.

3. Sugar industry utilises immobilised raffinase to digest raffinose present in sugar beet juice; this helps to increase sugar recovery by 3% and reduces the cost of disposal of molasses. Raffinase can be isolated from the mould *Mortierella vinacea*. Sugar industry also exploits immobilised invertase to produce invert sugar.

4. Pharmaceutical industry involves the use of immobilised penicillin amidase to synthesise penicillin and cephalosporin—the antibiotics exploited commercially to combat bacterial diseases.

5. Chemical industry utilises immobilised nitrilase to produce acrylamide from acrylnitrile. The enzyme nitrile hydratase (or nitrilase) is obtained from *Rhodococcus*.

6.10 BIOSENSORS

A biosensor is an analytical device that involves the combination of biological active material displaying characteristic specificity with chemical or electronic sensor to convert a biological compound into an electrical signal. These electric signals are amplified, interpreted and displayed to measure the concentration of compound present in the solution.

In design, the enzyme electrode or biosensor is composed of a given electrochemical sensor in close contact with a thin permeable enzyme membrane capable of reacting specifically with the given substrates. The embedded enzymes in the membrane produce O_2, hydrogen ions, ammonium ions, CO_2, heat, light or even directly electrons depending upon the enzymic reaction occurring, which are readily detected by the specific sensor. The magnitude of the response determines the concentration of the substrate (figure 6.5). Sensor can measure a reduction in one of the substrates or an increase in one of the products of the reaction catalysed by the biosensor. Although the biological component in a biosensor may more often be an enzyme or multi-enzyme system, it can also be an antibody, nucleic acid, an organelle, a microbial cell, whole slices of tissue or entire organs.

Considering an example of a simple glucose electrode which can be constructed by immobilising a layer of glucose oxidase, in polyacrylamide gel around a platinum oxygen electrode, the concept can be made clear. When a solution of glucose is brought into contact with the electrode, glucose and O_2 diffuse into the enzyme layer and are converted to gluconolactone and H_2O_2, lowering the O_2 concentration in the gel layer around the electrode. The rate of reaction is recorded as the rate of depleting O_2 concentration. Such a device responds linearly to glucose concentration over a range of 10^{-1}–10^{-5} mol dm^{-3} with a typical response time of 1 minute and is stable for up to 4 months.

The biosensor is of various types depending on the physical changes that occur in the vicinity of the sensor. These physical changes may be (i) heat released or absorbed by the reaction that can be measured by calorimetric biosensors, (ii) production of an electric potential due to altered distribution of electrons that can be detected by potentiometric biosensors, (iii) movement of electrons due to redox reaction which can be measured by amperometric biosensors, (iv) light produced or absorbed during reaction that is detected by optical biosensors, or (v) change in mass of biological component as a result of reaction, which can be measured by acoustic wave biosensors.

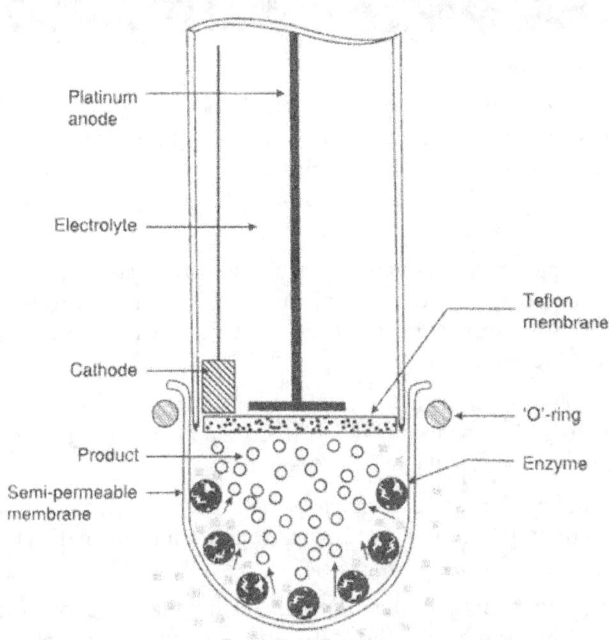

Figure 6.5   Biosensor developed with help of immobilised enzyme

Biosensors have tremendous role to play in world economy. The estimated world market is approximately $ 25 billion/year of which 30% is in healthcare. At present only 0.1% of this market is using biosensors. Biosensors have been constructed to measure almost anything from blood glucose level to the freshness of fish. Looking to the future, we can expect that biosensors will play an indespensable role in detecting pollutants present in the environment. Biosensor market is likely to flourish with the development of economical, easy-to-use and more stable biosensors.

Key Concepts

because of their enormous catalytic activity and high degree of specificity.

- Based on the type of reaction which they catalyse, enzymes are classified into six groups and named by the Commission on Biochemical Nomenclature as oxidoreductases, transferases, hydrolases, lyases, isomerases and ligases or synthetases.

- Enzymes may be isolated from almost any organism, but in practical terms, the use of microorganism, to produce enzymes has a number of technical and economic advantages.

- Genetic engineering for industrial production of an enzyme is possible by the transfer of the gene/s encoding useful enzymes into suitable host organism.

- Synzymes or artificial enzymes are synthetic polymers may be protein or nonprotein molecules having catalytic activities.

- Ribozymes are modified RNA molecules that have catalytic activity for breaking the RNA and have significant role to play in fight against the viral infections like AIDS and cancer, in addition to producing transgenic plants with increased viral resistance.

- Immobilisation of an enzyme is to fix the enzyme in some way onto a polymeric support matrix. Immobilisation alters the enzyme's behaviour in different ways.

- Soluble enzymes have tremendous applications in detergents, baking, brewing, dairy, textile food, drink, confectionary, leather, pharmaceutical and medical industries.

- Immobilised enzymes have significant role to play in food, dairy, sugar, pharmaceutical and chemical industries.

- Biosensor is an analytical device used to detect concentration of substrate in a solution.

- Biosensors have been constructed to measure almost anything from blood glucose level to the freshness of fish and even the pollutants. Thus they have significant application in analytical and clinical chemistry.

Reveiw Questions

1. What are enzymes? How are they classified?

2. Define enzyme technology. Explain the role of enzyme technology is solving the modern-day problems.

3. What are biological sources for enzymes? Explain how microbial enzymes are advantageous than plant or animal-derived enzymes.

4. Define enzyme engineering. Enlist the objectives of enzyme engineering with its principle.

5. Define enzyme immobilisation. What are the different methods of immobilisation?

6. Discuss the role of soluble enzymes and immobilised enzymes in different industrial applications.

7. What is biosensor? Explain its principle with its applications.

8. Write short notes on:

    a. Enzyme induction and repression

    b. Extremozymes

    c. rDNA technology for enzyme production

    d. Synzyme

    e. Role of Ribozyme in viral infection

    f. Encapsulation

    g. Enzymes in medicine

    h. Biosensor

    i. Bating

7

SINGLE CELL PROTEIN

7.1 INTRODUCTION

The explosive rate of population growth is the major problem faced by developing nations. In India the rate of population growth is far greater than food production and supply from all sources available to the population. Conventional agricultural practices may be unable to provide optimum food to the ever-increasing mouths and this results in shortage of proteinaceous food supply. The Food and Agriculture Organization (FAO) predicted a widening of protein gap between developed and developing countries. At least 25% of the world's population currently suffers from hunger, malnutrition and starvation. Due to these consequences the population becomes susceptible to various types of diseases.

Biotechnological innovations will accelerate the trend of food productivity in all branches of agriculture. However, there are still major imbalances in availability of cereals; there are further disturbances in production of food due to alterations in global weather patterns (which lead to unseasonable rainfall), national and international warfare, arid and drought conditions and infertile land.

Biotechnologists are searching for sources of protein. New agricultural practices are making efforts to develop high protein cereals. The cultivation of soybeans and groundnuts is ever-expanding, protein may be extracted from liquid wastes by ultra filtration, and the use of microorganisms as the producers of protein has gained experimental success. This area of study is called *single cell protein* or *SCP production*.

Professor Scrimshaw (1960) of MIT, USA introduced the term *single cell protein (SCP)* for microbial protein that includes the proteins from unicellular (single-celled) microorganisms like bacteria yeasts, algae and possibly protozoa. In addition, mushrooms which are fruiting bodies of certain basidiomycetous fungi are also rich sources of protein. Some of the mushrooms are non-poisonous and edible due to their pleasant flavour and richness in protein, and several major companies throughout the world have long been actively involved in SCP production and mushroom cultivation on a commercial scale.

7.2 MICROORGANISMS—SOURCE FOR SCP

Algae, fungi, yeast and bacteria are the microorganisms from which SCP is produced. The microbes utilised for SCP production must be nontoxic and nonpathogenic to plants, animals and man. They should be fast growing, easy to separate from the medium and easy drying. Further they must have good nutritional value and must be able to be cheaply produced on a commercially large scale.

Microorganisms produce protein much more efficiently than any other animal. They can grow at remarkably rapid rates under optimum conditions; some microbes can double their mass every 0.5 to 1 hour. The protein-producing capacities of 250 kg cow and 250 g of microorganisms are often compared. The cow can produce 200 g of protein per day, whereas the microbes theoretically could produce 25 tonnes in the same time under growing conditions. However, the cow has its own significance because of its unique ability to convert grass into protein rich milk. After several years of research no rival method for that conversion process has been developed. Therefore cow has been rightly described as a bioreactor. The use of microorganisms for SCP production is also advantageous because they are more easily genetically modified using gene transfer technology than plants and animals. In addition, they can grow on a wide range of raw materials.

7.2.1 SCP from Algae

The use of algae as SCP producer is gaining interest because they grow well in ponds or tanks and need only the freely available CO_2 as a carbon source and sunlight as an energy source for photosynthesis. Members of the genera *Chlorella* and *Senedesmus* have long been used as food in Japan, while *Spirulina* is widely used in Africa and Mexico. *Spirulina maxima* is commercially produced in Mexico as a by-product of a large solar evaporator used for production of soda lime. It is used as animal feed and also suited for human use. It may be harvested by filtration or simply by skimming. *Chlorella* is used as protein and vitamin supplement in some Japanese yoghurt, ice cream and breads. Algae can be used in ponds or lagoons to help in the removal of organic pollutants and the resultant biomass is harvested, dried and the powder added to animal feed. Algal SCP contains about 60% crude protein that increases their value as animal feed; the only disadvantage is its rich chlorophyll content that makes it unsuitable for human use.

7.2.2 SCP from Filamentous Fungi

Pekilo, a fungal product that was produced by fermentation of carbohydrates derived from sulphite liquor molasses, whey, waste fruits, woods, starch hydrolysates or agricultural hydrolysates, contains about 55% protein and is rich in vitamins.

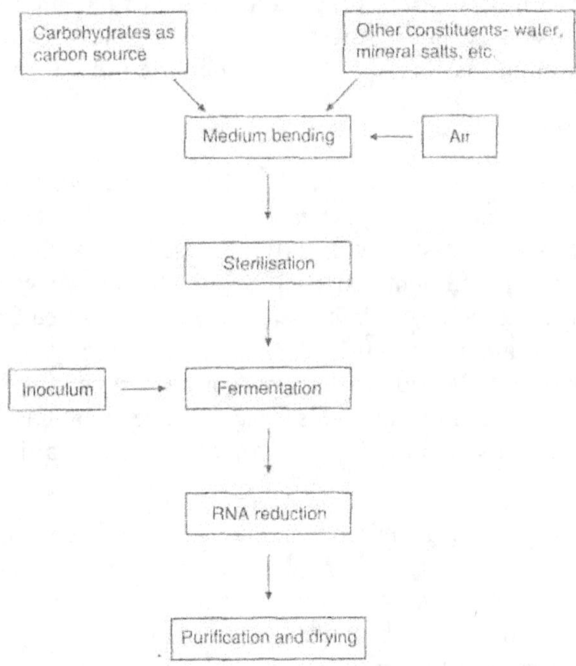

Figure 7.1   Flow diagram of mycoprotein production

The filamentous fungus, *Paecilomyces variotti*, is used in continuous fermentation processes to produce Pekilo, which is a good protein and vitamin supplement in the diet of farm animals and poultry birds. In UK, Rank-Hovis, Mc Dougall in conjunction with ICI (Marlow foods) is now commercially marketing the fungal protein, mycoprotein or Quorn, derived from the growth of *Fusarium graminearum* fungus on simple carbohydrates. The various stages for the production of mycoprotein from this fungal strain are shown in figure 7.1. Since they have high nucleic acid content (about 15% RNA, acting as the source of non-protein nitrogen), the process of continuous culture of the microorganism involves the reduction of RNA that is achieved by addition of RNase (ribonuclease). Towersey, Logton and Cockram have patented this process in 1976. The process increases the actual percentage of protein in the reaction mixture.

Unlike almost all other forms of SCP, mycoprotein is produced for human consumption. The product is open to be sold in the UK since it is nontoxic and compatible with normal physiological function. In studies of its routine toxicity to man, one of the important physiological effects observed was a reduction in serum cholesterol.

7.2.3 SCP from Yeast

The commercial production of SCP from *Candida utilis* (Torula yeast), *C. intermedia*, *Kluyveromyces fragilis*, and *Saccharomyces cerevisiae* (Baker's yeast) is currently practised using the substrates like confectionary effluents, ethanol, sulphite liquor, whey, n-alkanes (C_{10}–C_{23} hydrocarbons), molasses, etc. The SCP produced from yeast has protein content upto 60%. The first commercial production of SCP dates back to World War I when Torula yeast was produced in Germany and used in soups and sausages. Since then SCP produced from yeast is used both for human food and in animal food supplement.

Members of yeast also have high nucleic acid content which need to be reduced by treating them with RNase; further, their slower growth rate may produce difficulties in production of SCP. However, there is minimum risk of bacterial contamination and their recovery by continuous centrifugation is easy.

7.2.4 Bacteria as Source of SCP

Several bacterial strains have been utilised as SCP producers. The Imperial Chemical Works (U.K.) produces pruteen, the SCP product of *Methylophilus methylotrophus*, a bacterium that grows on C_1 compounds. This bacterium is grown on methanol derived from methane and the cell crop is harvested, centrifuged, dried and sold in pellet or granular form. Another bacterium *Brevibacterium* sp. is also exploited for its SCP-producing ability that uses C_1–C_4 hydrocarbons as a substrate. Bacterial SCP has very high and crude protein (up to

80%), which add to its nutritional value but at the same time it has high RNA content (upto 20%), which is to be reduced; it is prone to contamination by pathogenic bacteria. Though their growth rate is highest and they use wide range in substrate, their recovery process is problematic.

Key Concepts

- Traditional agricultural practices are unable to supply protein-rich food to the ever-increasing population. With the advent of biotechnology, wide experimental success has been gained in using microorganisms as producers of proteins.

- The protein derived from unicellular microorganisms like bacteria, yeast, filamentous fungi and algae is called single cell protein (SCP).

- SCP produced from algal members such as *Chlorella, Senedesmus* and *Spirulina* is widely used as animal feed and also suited for human use.

- Algal SCP contains 60% of crude protein while that from filamentous fungi contains 55% protein and is rich in vitamin.

- *Paecilomyces variotti* is a filamentous fungus used to produce Pekilo—the protein and vitamin supplement produced on a commercial basis and employed in the diet of farm animals and poultry birds. Quorn or mycoprotein is derived from *Fusarium graminearum* using simple carbohydrates as a substrate.

- *Candida utilis* (Torula yeast), *Kluyveromyces fragilis* and *Saccharomyces cerevisiae* (Baker's yeast) play significant role in industrial production of SCP using the substrates like sulphite liquor, whey, molasses, n-alkanes, etc. SCP produced by yeast has protein content upto 60% that make it suitable for human food and animal food supplement.

- The bacterial strain *Methylophilus methylotrophus* produces SCP called pruteen which has up to 80% of crude protein significantly exploited for human and animal feed.

7.3 SCP DERIVED FROM FOSSIL CARBON SOURCES

The substrates having high commercial value as energy sources or derivatives of such chemicals, e.g. gaseous hydrocarbons, liquid hydrocarbon, gas oil, n-alkenes, methanol and ethanol have wide commercial importance in SCP production.

Methane (C_1 - hydrocarbon) as an SCP source has been extensively studied but there are technical difficulties, like risk of explosion when more than 12% O_2 (v/v) is used and requirement of efficient cooling during the process which hampered the production. In contrast, methanol offers great interest. In UK, ICI constructed a large-scale fermentation plant for producing the methanol-utilising bacterium—*Methylophilus methylotrophus.* Hoechst (West Germany) and Mitsubishi (Japan) worked on a process—they used yeast as the SCP source instead of bacteria. Methanol as a carbon source for SCP has many advantages over n-paraffins, methane and even polysaccharides, since methanol is nontoxic, easily soluble in water, and its composition is independent of seasonal functions. C_5 to C_8 n-alkanes are liquid at room temperature and they are usually toxic to cells due to their solvent action. Generally C_9–C_{18} n-alkenes are used as SCP substrates. British Petroleum was producing SCP at an industrial scale using gas oil in combination with ammonia and mineral nutrients with the help of *Candida* yeast. Ethanol is also acceptable as a substrate for SCP production for human use. Bacteria, yeast and filamentous fungi exploit ethanol for SCP production. In USA, Amco Foods produces SCP by growing *Torula* yeast.

7.4 SCP FROM WASTES

Agricultural, forestry and industrial wastes like straw, bagasse, sawdust, sulphite liquor, effluents from different industries, starch and cellulose hydrolysate, whey, molasses, animal manures and sewage, significantly contribute to pollution of water bodies. The vitalisation of such materials for SCP production offers two benefits—reduction in level of environmental pollution and production of edible protein. Most of the organic wastes are available at low cost in most countries thus ensuring independence in supply. Molasses acted upon by Baker's and *Torula* yeast, whey by *Kluyveromyces fragilis* in combination with *Lactobacillus bulgaricus*, cellulosic material by fungus *Chaetomium cellulolyticum*, starch hydrolysate by *Fusarium graminearum,* and industrial effluent by *Candida utilis* and *Paecilomyces varioti* can produce SCP of human interest.

7.5 MUSHROOMS AS A SOURCE OF PROTEIN

Mushrooms are fruiting bodies of certain basidiomycetous fungi. Edible mushrooms with pleasant flavour and high protein content are eaten directly as food in the form of soup, curry or with meat. A wide range of edible mushrooms is cultivated throughout the world (table 7.1) for human consumption. Basidiomycetes or mushroom type fungi have the ability to degrade lignocellulose material like wood, straws, sawdust and rice bran. The cultivation of edible mushrooms is one of the examples of a microbial culture in which the cultivated organism itself that is a macroscopic, highly developed umbrella-like structure that can be directly utilised as human food. The production of mushroom is in principle a fermentation process where the lignocellulose material is composted with other suitable ingredients (like animal manures and other organic nitrogen compounds) over a specific time and the final product is a unique substrate utilised by the spores present in the inoculum (spawn) to develop into edible mushroom under certain requisite conditions that includes sterilisation, addition of insecticides, maintenance of temperature and humidity, etc. On a worldwide basis, mushroom cultivation is one of the fastest growing biotechnological industries and it is expected that it will expand even further with the production of enzymes and pharmaceutical compounds such as antibiotics and anticancerous drugs. The production of mushroom carries several advantages; few of them are as follows:

   i. Their process of production is simple and easy.

  ii. Low cost substrates are utilised for their production.

 iii. Edible mushrooms are rich sources of protein.

 iv. They also have B-complex vitamins that are well-preserved during drying, cooking, etc.

  v. During the process of cultivation of mushrooms, the plant residues and waste lignins are converted to useful edible material.

 vi. Mushrooms protect the environment from lignin-derived pollutants.

Table 7.1   Some cultivable edible mushrooms

Species	Local name	Substrate	Country
Agaricus bisporus	Button mushroom	Horse manure, wheat, rice straw	India, Europe, U.S.A.
Auricularia sp.	Woodear	Rice bran, saw dust	U.S.A.
Coprinus sp.	-	Straw	-
Flammulina velutipes	Winter mushroom	Sawdust, straw	Europe, U.S.A
Grifola frondosa	Sitting hen mushroom or Limuo	Straw	Europe, U.S.A.
Hericium erinaceus	Monkey head or hedgehog mushroom	Wheat/rice straw	U.S.A.
Hypsizigus marmorens	Shimeji	Wheat/rice straw	Asia
Lentinula edodes	Shiitake or oak mushroom	Log, paper saw dust	U.S.A., Asia, Europe
Pleurotus ostreatus	Oyster mushroom	Straw, rice bran, saw dust	India, China Japan, U.S.A. Europe
Pholiota nameko	Nameko or viscid mushroom	Wheat/rice straw	Asia, Europe
Termelia fuciformis	White jelly fungus or 'silver ear'	Wheat straw, rice bran	Asia, U.S.A.
Volvariella volvacea	Padi-straw mushroom or Chinese mushroom	Paddy/wheat/rye straw, cotton	Asia

Key Concepts

- SCP is also derived from fossil carbon sources like gaseous or liquid hydrocarbons, gas oil, n-alkanes, methanol and ethanol using several bacteria, yeast and filamentous fungi.

- Agricultural and industrial wastes are also degraded by yeast and fungal strains to produce SCP of human interest.

- Cultivation of edible mushrooms like *Agaricus bisporus* (Button mushroom), *Pleurotus ostreatus* (Oyster mushroom) and *Volvariella volvacea* (Padi straw mushroom or Chinese mushroom) on low-cost substrates like plant residues and waste lignins, is one of the fastest growing biotechnological industries. Edible mushrooms are rich source of protein (about 51% on dry weight basis).

- SCP product, before releasing into market, must be assessed for its physical, chemical and toxicological features.

- The acceptability of SCP as human food depends on its safety and nutritional value. In addition, eating food derived from microbes has many psychological, social logical and religious implications.

- SCP production and mushroom cultivation provide great advantage in converting different wastes into edible products that can narrow the gap between demand and supply of quality protein.

7.6 SAFETY AND NUTRITIONAL EVALUATIONS OF SCP

A tremendous amount of attention has been paid to the problems related to the safety, nutritional value and acceptability of SCP.

1. The chemical assessment of SCP in terms of protein, nucleic acid, lipid and vitamin content must be done.

2. The nature of raw material used in SCP process must be analysed, e.g. the possibility of carcinogenic hydrocarbons in gas oil or n-alkenes, the heavy metals or other contaminations in the mineral salts and the mycotoxin production by certain fungi.

3. The process organism must be non-pathogenic and non toxigenic.

4. Rigorous sanitation and quality control procedures must be maintained throughout the process to avoid spoilage or contamination by pathogenic or toxigenic microbes.

5. The physical characteristics like texture, odour, taste, colour, particle size, density, storage, etc. must be determined.

6. Toxicological testing of final product including short-term acute toxicity must be done with different laboratory animal species.

7. For the evaluation of nutritional value of the product, long-term testing should be done on target species.

Thus, if SCP is to be used directly as food for human beings, the skills of the food technologist will be greatly challenged.

7.7 ECONOMIC IMPORTANCE OF SCP

The process of SCP production, the microorganisms involved in the process and the product itself offer several advantages which includes the following:

1. SCP is rich in high-quality protein, with poor quantity of lipids/fats which make it desirable for use as human food supplement.

2. The SCP may be used in animal feed to at least partially replace the currently used protein-rich soybean meal and fish-proteins and even cereals that can be diverted for human consumption.

3. The SCP may be a good source of vitamins particularly B-complex which adds to its nutritional value.

4. SCP can be stored and shipped over long distance without spoilage.

5. The process of production can be continued throughout the year independent of the seasonal fluctuations.

6. The microorganisms involved in the production process have remarkably fast-growing ability and they can produce large quantity of SCP in a relatively small space.

7. The microbes can easily be modified genetically by gene transfer technology to have plenty of quality protein.

8. The substrates used in the process are of low cost and the effluents produced are easily degraded by microbes to reduce pollution.

9. Mushroom cultivation provides a great advantage in converting industrial, urban and wood wastes into a product that is directly edible by human beings.

10. The production of SCP and mushroom cultivation have evolved with new technical solutions to fill the gap between demand and supply of quality protein.

Review Questions

1. What is single cell protein? Describe the role of microorganisms in the production of SCP.

2. Explain the significance of microorganisms in the production of SCP.

3. Define mycoprotein. Draw a flow diagram of mycoprotein production using filamentous fungi.

4. Explain the role of microorganisms in the production of SCP using fossil carbon sources and organic wastes.

5. What are mushrooms? What are the different advantages offered by mushroom cultivation with different types of cultivable edible mushrooms?

6. Write short notes on:
    a. RNA reduction
    b. Spirulina
    c. Pekilo
    d. Padi mushroom
    e. Methanol as SCP substrate
    f. Nutritional and safety evaluation of SCP
    g. Economic importance of SCP
    h. Baker's yeast
    i. Role of mushrooms in reduction of pollutions

8

FERMENTATION TECHNOLOGY

8.1 INTRODUCTION

The term fermentation covers any process by which microbes are grown in large quantities to produce any type of product or service of economic value. The word bioprocessing, fermentation and industrial microbiology are virtually similar in their scope, objectives and activities. Bioprocess technology was derived in part from the use of microorganisms for the production of food, chemicals, pharmaceuticals, energy, agricultural products, etc. which are of great economic importance and which are exploited for human welfare (table 8.1). Parallel to these useful product formations, microorganisms also play a role in removing obnoxious and hazardous wastes, and this has resulted in the worldwide service industries involved in detoxification of waste and toxic compounds, water purification, effluent treatment and solid waste management.

Fermentation involves a multitude of complex enzyme-catalysed reactions using specific microorganisms under critical physical and chemical conditions. Traditionally, fermentation processes were employed for production of wines, bread, etc. Still these practices are in use and with the advent of technology new commercially significant bio-products are increasingly being derived from microbial fermentation, and include:

i. Primary metabolites, e.g. acetic acids, lactic acids, glycerol, acetone, butyl alcohol, amino acids, vitamins and polysaccharides.

ii. Secondary metabolites (that are produced as by-products of cellular metabolism by microbes and are not essential for the survival and growth of organism) namely, antibiotics, growth hormones, inhibitors, tannins, resins, alkaloids, steroids, etc.

iii. Industrially useful enzymes, e.g. amylases, pectinases, proteases, invertases, asparaginase, restriction enzymes, etc.

iv. Products obtained by fermentation of various substrates using yeast, algae, filamentous fungi and bacteria that are collectively called biomass (e.g. single cell protein) and that are employed as food, feed and fodder.

All these microbial products have commercial importance and are essential to modern society. Recently, bioprocess technology is increasingly using cells derived from plants and animals to produce several significant bio-products such as the following.

Table 8.1   Industrial sectors and microbial products/activities for commercial use

Industry	Products/activities
1. Chemicals	
a. Organic solvents	Ethanol, butanol, acetone
b. Organic acids	Lactic acid, citric acid, acetic acid
c. Inorganic	Bioaccumulation and leaching of metals
d. Gases	CO_2, H_2, CH_4
e. Amino acids	L-glutamic acid, L-lysine
2. Pharmaceuticals	
a. Antibiotics	Penicillin, streptomycin, tetracycline
b. Vitamins	B_{12}, riboflavin, A
c. Recombinant proteins	Interferon, insulin, subunit vaccines, growth hormone
d. Others	Enzyme inhibitors, steroids, vaccines
3. Food	
a. Dairy products	Cheese, pickles, yoghurt, bread, vinegar
b. Beverages	Wine, beer, whisky, tea, coffee
c. Food additives	Antioxidants, colour, flavours, stabilisers
4. Biomass	SCP and mushroom products
5. Energy	Ethanol (gasohol), Methane (biogas) and Biomass
6. Agriculture	Microbial pesticides, biofertilisers, mycorrhyzae, Rhizobium and other N-fixing bacterial inoculants, veterinary vaccines, plant cell and tissue culture
7. Petroleum	Enhanced recovery of oil and discovery of new oil reserves
8. Environmental wastes,	Disposal of biological and industrial solid waste treatment by composting waste water and sewage treatment by anaerobic digestion

i. Plant cell and tissue culture involving vegetative propagation, embryogenesis and genetic improvement is aimed largely at secondary products formation such as flavours, perfumes and drugs.

ii. Mammalian cell culture has been concerned with production of vaccine, antibodies and pharmaceutical proteins such as interferon, interleukins, etc.

Fermentation technology essentially involves the use of suitable microorganisms, nutrient medium, optimum temperature and suitable pH and above all a physical or technical containment system called fermenter or bioreactor. In the bioreactor the chemical changes are brought about in an organic compound by the action of enzymes secreted by microorganisms under suitable physical and chemical conditions to produce a commercial product on a large scale.

8.2 FERMENTER OR BIOREACTOR

The biotechnological process is carried out in a specially designed container where the correct environment for optimisation of growth and metabolic activity can be possible to produce the commercial product. That container is called fermenter. In simple words, the fermenter is a container where fermentation is carried out; it may be a large vessel constructed as an upright cylinder. It is usually made up of stainless steel or copper of suitable grade to provide resistance to steam sterilisation. Majority of fermenters are fitted with stirring gear for agitation that helps in breaking clumps of organisms. Bioreactors range from simple stirred or non-stirred open containers to complex aseptic, integrated systems with varying levels of advanced computer inputs (figure 8.1).

Fermenters are of two distinct types. The first is open that allows continuous processing with substrate entering at one end and products leaving at another. It is primarily a non-aseptic system where it is not absolutely essential to operate with entirely pure cultures, e.g. brewing, effluent disposal systems. In the second closed type the processing is done in batches and aseptic conditions are essential for successful product formation. e.g. antibiotics, vitamins, etc.

The microbes are grown on nutrients placed in a typical fermentation vessel at the start of the fermentation. The vessel is cooled by a water jacket, filtered air is pumped into the bottom of the liquid, and acid or alkali added as necessary. A stirrer keeps the contents well-mixed. Steam lines are provided so that the container can be sterilised after each fermentation batch.

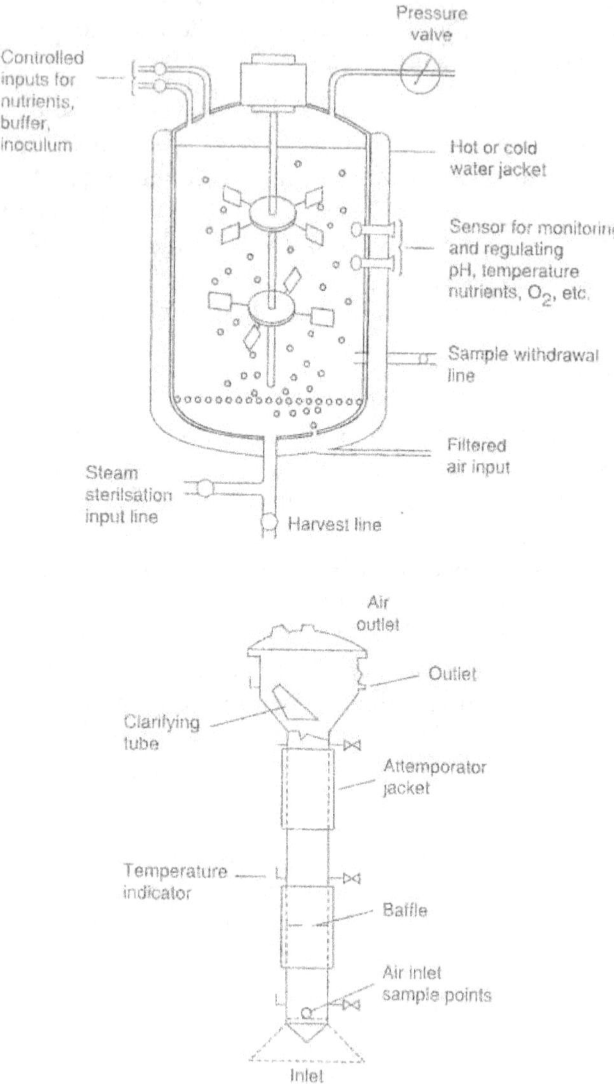

Figure 8.1   (a) Continuous stirred tank fermenter   (b) Tubular tower fermenter

For designing the fermenter and to achieve optimisation of the bioreactor system, the following points should be considered:

i. The fermenter should provide an environment suitable for the growth of a pure culture and/or a defined mixed culture, which can run free from contamination by unwanted microbes.

ii. The culture volume inside the fermenter should remain constant i.e. it is essential to avoid undue release of culture organism into environment. Indeed, it is very essential to retain the organisms within the bioreactor particularly when genetically engineered microbes are a part of culture.

iii. All the nutrients, including dissolved oxygen must be provided to diffuse into each cell, and waste products like heat, CO_2 and waste metabolites should be removed.

iv. During fermentation, it is necessary to regulate many factors within predetermined values: oxygen, CO_2, pH, temperature, nutrient media concentration, etc.

v. The standard of materials used in the designing of bioreactors is also important i.e. the materials must be corrosion-resistant, nontoxic, be able to withstand repeated sterilisation with high pressure steam and should not be deformed or broken under mechanical stress. If possible the material should be transparent since visual inspection of the medium and culture is advantageous.

8.3 STEPS INVOLVED IN FERMENTATION PROCESS

The fermentation process is usually developed in three stages. In the initial stage basic screening procedures are carried out by relatively simple microbiological techniques, using petri dishes, Erlenmeyer flasks, etc. This is followed by a pilot plant investigation to determine the optimal operating condition in a volume of 5 to 200 litres. The third stage is the transfer of technology to plant or production scale and final economical exploitation of the desired product.

The general steps required to optimise fermentation are shown in figure 8.2 and are as follows:

1. Isolation and screening of microorganism The success of fermentation mainly depends on the microorganism strain used. The strain that is genetically pure, easy to manipulate by genetic engineering, and having minimum risk of contamination should be selected. Industrially important microbes like bacteria, fungi, algae, etc. are isolated from soil, lakes and river mud. The desired microbes are then screened by studying their products or biosynthetic pathways and a pure culture of selected microbes is developed.

2. Selection of nutrient medium   The fermentation medium includes nutrients essential to achieve optimum growth of microbes and formation of desired end products. The choice of the nutrient is based on nutritive value and economic aspects. Usually locally available and cheap raw material like corn sugar, molasses, starch, vegetable oils, soybean meal, corn steep liquor, etc. is selected for preparation of nutrient medium.

3. Pretreatment   The selected raw material is treated and made suitable for fermentation. It may involve hydrolysis, dilution, acidification, etc. For example molasses are treated for removal of iron salts in it. Similarly, starch in corn syrup is hydrolysed to sugar before yeast can convert it to ethanol.

4. Sterilisation   The fermenter and nutrient media are sterilised to inactivate or to remove the living organisms and can be done by heating, irradiation, filtration and treatment with chemicals like formaldehyde, H_2O_2, etc. This avoids contamination of fermenter and nutrient media by unwanted microorganisms.

5. Transfer of nutrient medium   The pretreated and sterilised nutrient medium is transferred to the fermenter for the process.

6. Inoculation   The medium is inoculated with pure culture of selected microorganisms to achieve fermentation.

7. Adjustment of pH and temperature   The nutrient medium is adjusted with optimum pH and temperature so that the activities of microorganisms can produce desired products in sufficient quantity. At the same time medium requirement, aeration, agitation and duration should also be adjusted as per predetermined values.

8. Recovery of products   The processes used for actual recovery of the desired product after completion of fermentation is called downstream processing. Since the desired product may be present in very small quantities, it is generally mixed with other molecules from which it has to be separated. The downstream processing involves filtration, centrifugation, flocculation and floatation to separate the cells from liquid medium. Various methods like extraction, concentration and purification are employed to disintegrate the cells and finally the product is dried, making it suitable for handling and storage. In many cases, the fermentation products are inhibitory to the product formation. In such situation, *in situ* recovery of the products is employed to increase the yield of the desired product. This can be achieved by use of vacuum, adsorbant, anion exchange resin or dialysis.

Once the fermentation process is completed and the product is recovered, it is necessary to clean the bioreactor.

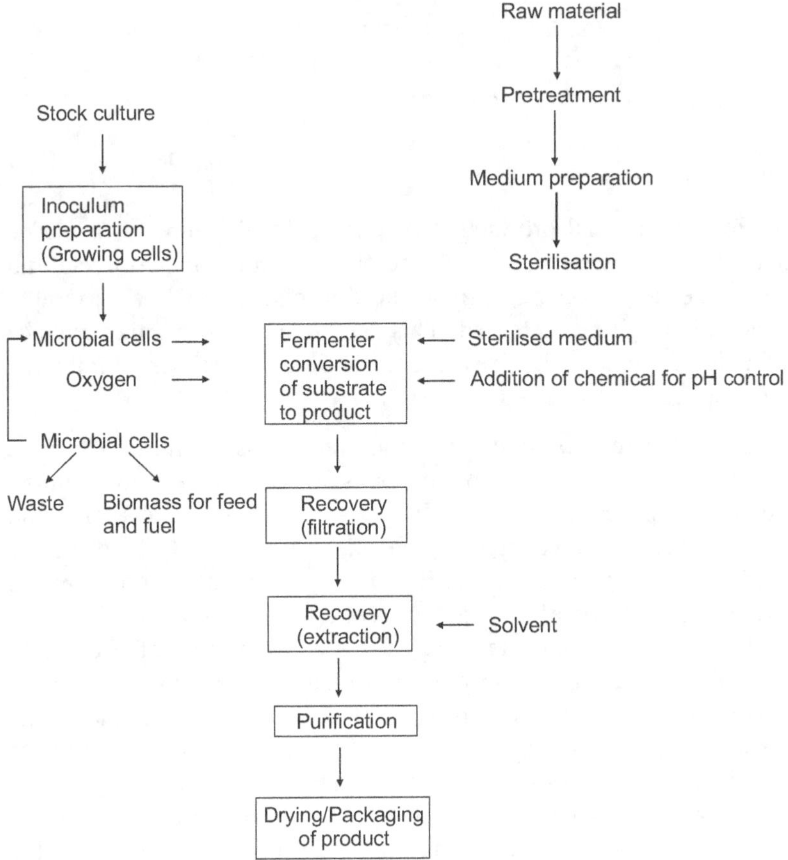

Figure 8.2   Steps involved in the process of fermentation

8.4 PRODUCTION OF CITRIC ACID BY FERMENTATION

Citric acid is the fermentation product of numerous microbes, however the culture of fungus strain *Aspergillus niger* is used to produce commercially high yield of citric acid.

In the Kreb's cycle, citric acid is the first principal organic acid hence it is also called citric acid (CA) cycle. When the mycelium of *Aspergillus niger* grows, CA is produced as an intermediate in CA cycle and is further directed to growth-promoting biosynthesis or energy release as shown below.

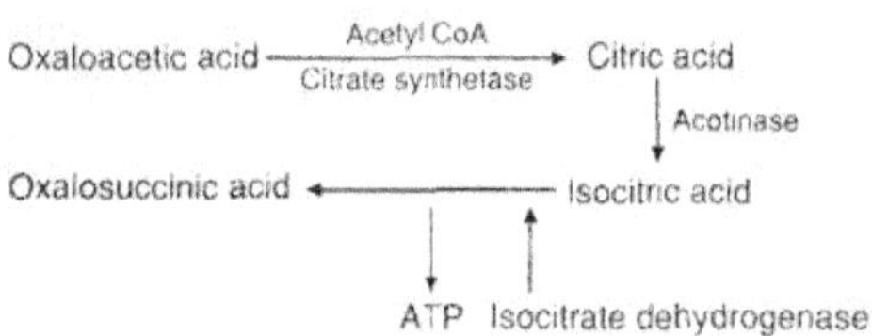

For commercial production of citric acid, the activity of citrate synthetase is to be increased and the activities of acotinase and isocitrate dehydrogenase are to be diminished so that citric acid production will be increased. This can be achieved by maintenance of culture media at low pH and by addition of enzyme inhibitors, e.g. copper ion.

In commercial production of citric acid, *Aspergillus niger* may be cultured on the surface of shallow pans or as submerged mycelium in aerated vats. Beet molasses is the low-cost carbohydrate source and has high metallic ion content which may cause reduced yields of citric acid. These metallic cations are removed by the use of cation-exchange processes and adsorbants or by treatment with ferricyanide. The carbohydrate source is then diluted to a concentration of about 20% to 25%, since a higher concentration inhibits the formation of acids other than citric acid. The process of fermentation to manufacture citric acid is adjusted to low pH , and HCl is added to the medium to the range of pH 2 to 5 when the fungal spores in the inoculum germinate. Nitrogen is provided by addition of ammonium salts. Copper may be added as antagonists to iron and this helps in the control of mycelia growth to increase citric acid production. The prepared medium is added with fungal spores from stock cultures.

Highly aerobic conditions are required, and submerged cultures are aerated with sterile air and agitated during the incubation period that may last for 7 to 10 days. The temperature of the nutrient medium is maintained in the range of 25–30 °C. Finally harvesting of citric acid is started. Lime is added to the culture medium to precipitate any oxalic acid formed during the process. Mycelium and calcium oxalate are filtered off. The slurry of calcium citrate is precipitated out. It is then treated with sulphuric acid to precipitate calcium. The dilute citric acid solution is purified by treatment with carbon and is demineralised by successive passage through cation- and anion-exchange resins. This purified citric acid solution is then evaporated and crystals of citric acid are obtained which are further purified by recrystallisation process using water.

Citric acid thus produced is commercially exploited as a flavouring agent in beverages and foods such as soft drinks, desserts, jams, jellies, candies, wines and frozen fruits. The acid is rapidly and almost completely metabolised in the human body and has wide pharmaceutical uses. Its anticoagulating property is of immense importance in the preservation of components of whole blood playing an indispensable role in blood transfusion. Citric acid is also used in astringent lotions to adjust the pH, in hair rinses and hair setting preparation, in electroplating, in leather tanning, etc. Citric acid has also role to play as a metal-chelating and sequestering agent and as a plasticiser.

8.5 LACTIC ACID FERMENTATION

The commercial production of lactic acid utilises bacterial strains like *Lactobacillus delbrueckii, Streptococcus* sp. and *Leuconostoc* sp. The nutrient medium for the production of lactic acid contains 10–15% glucose, 10% calcium carbonate to neutralise the lactic acid formed, ammonium sulphate and trace amounts of other nitrogen sources. The carbohydrate sources used in the fermentation process are corn sugar, beet molasses, potato starch and whey.

The temperature of the medium is kept at 45–50 °C while the pH is adjusted to 5.5–6.5. There is no need of aeration since the fermentation process is carried out anaerobically. The process of fermentation is completed within 5–7 days. Calcium carbonate is added after fermentation to raise the pH to 10. The medium is heated and filtered and then lactic acid is recovered. During the process of fermentation, lactic acid can also be recovered by dialysis in which the fermenter may contain a dialysis chamber separated by a selectively permeable membrane. The lactic acid diffuses from the fermentation chamber into the dialysis chamber (by one-way diffusion) and from this chamber, lactic acid is recovered while the fermentation continues.

Lactic acid is used as a food preservative, for deliming hides in leather production and for fabric treatment in textile industry. Lactic acid derivatives are utilised for various purposes—polylactic acid in resins, copper lactate in electroplating, calcium lactate in baking powder and animal feed supplement.

8.6 PRODUCTION OF VITAMINS BY FERMENTATION

Vitamins are organic compounds that are essential for the normal health and complete growth of organisms. They are naturally synthesised by green plants and microorganisms while animals including man cannot synthesise vitamins hence they must be included in diet. The absence of vitamins causes deficiency diseases like night blindness, beriberi, scurvy, rickets, etc. Several vitamins can be produced by microbial fermentation. A list of microorganisms producing specific vitamins, the medium and the fermentation condition for their growth is given in table 8.2.

Table 8.2   Some of the vitamins produced by microbial fermentation

Vitamins	Microbial culture	Medium	Fermentation condition
Vitamin B_2 (Riboflavin)	Fungus *Ashbya gossypii*	Glucose, Collagen, Corn steep liquor, Soya oil, Glycine.	pH 6.0 to 7.5 4 to 5 days at 28–30 °C Aerobic
Vitamin B_{12} (Cobalamine)	Bacteria- *Pseudomonas denitrificans*	Sucrose, Cobalt chloride,	pH 6 to 7.7 2 days at 26–28 °C
	Streptomyces griseus	Betaine, salts	Aerobic

Table 8.2   *Contd...*

Vitamins	Microbial culture	Medium	Fermentation condition
		Glutamic acid, 5,6-dimethyl–benzimidazole.	
Vitamin B$_{12}$ (Cobalamine)	*Propionibacterium shermanii*	Dextrose, Corn steep liquor, Ammonia,	pH 7 3 days at 30 °C Anaerobic + 4 days
Aerobic		Cobalt.	
L-sorbose (in vitamin C synthesis)	*Gluconobacter oxidans*	D-sorbitol, 30% corn, steep liquor	pH 7, 45 hours at 30 °C Aerobic
5-keto gluconic acid (in vitamin C synthesis)	*Gluconobacter oxidans*	Glucose, CaCO$_3$, Corn steep liquor	pH 7 33 hr at 30 °C Aerobic

Vitamins produced by large-scale fermentation are very essential for the human society. Vitamin B-complex includes vitamins like B$_1$, B$_2$, B$_6$, and B$_{12}$. It prevents beriberi (a type of paralysis caused by degeneration of nerves) and anaemia. It also stimulates growth and appetite. Vitamin C prevents scurvy—a disease of the gums and teeth.

Key Concepts

- Fermentation is a biotechnological process in which microorganisms bring about changes in the raw material to arrive at a product of commercial importance.

- Industries like chemical, pharmaceutical, food, biomass, energy, agriculture, petroleum, etc. derive benefits from fermentation or bioprocess.

- Even waste water and sewage treatment, and disposal of industrial as well as biological wastes depend on bioprocess technology.

> ### Key Concepts
>
> ❑ Commercial production of fermentation products involves the following general steps
>
> > 1. Isolation and screening of microorganisms
> >
> > 2. Selection of nutrient medium
> >
> > 3. Pretreatment of raw material
> >
> > 4. Sterilisation of fermenter
> >
> > 5. Transfer of nutrient media into fermenter
> >
> > 6. Inoculation
> >
> > 7. Adjustment of pH and temperature for optimisation of fermentation
> >
> > 8. Recovery of product by downstream processing
>
> ❑ The fungal strain- *Aspergillus niger* is used for commercial production of citric acid, the bacterial strain *Lactobacillus delbrueckii* is employed for lactic acid fermentation.
>
> ❑ Citric acid, lactic acid and vitamins produced during fermentation have immense significance in modern life.

Review Questions

1. Define fermentation. Explain the role of fermentation technology in human welfare.

2. Enlist the industrial sectors exploiting fermentation technology.

3. What is a fermenter or bioreactor? What are the requisite conditions for the designing of a fermenter?

4. Explain various steps involved in the general process of fermentation with the help of a flow diagram.

5. Describe the principle and method of citric acid fermentation. Give the uses of citric acid.

6. Write short notes on:

    a. Vitamin B_2 and B_{12}

    b. Secondary metabolites

    c. Downstream processing

    d. Lactic acid fermentation

    e. Vitamin C

9

BIOFUEL TECHNOLOGY

9.1 INTRODUCTION

Throughout the world 'energy crisis' is a cause of great concern because of the rapidly declining fossil fuel reserves. Modern industries are almost totally dependent on the energy derived from the limited fossil fuel mainly coal, natural gas, and oil. The crux of the problem is that we do not yet have the right technologies to tap much of the energy present in our surrounding in the form of biomass and sunlight. *Biofuels* are the fuels generated from the biomass with the help of biological agents; these are namely bioethanol, biomethane, biobutanol and biodiesel. Biofuels are considered as substitutes for the rapidly depleting, nonrenewable and increasingly expensive fossil fuels. Biomass is the organic material of biological origin such as agricultural, forest and animal residues and industrial and domestic wastes that can be converted by physico-chemical and/or fermentation processes to produce pollution-free, cheap and ample **biofuel**. All these processes together constitute **biofuel technology**. The technology generating biofuel ultimately serves the nation, the world and this planet to save it not only from various types of pollution but also from the crippled energy situation.

9.2 SOURCES OF BIOMASS AND THEIR MODE OF UTILISATION

Photosynthetically derived biomass exists in many available forms in the environment that can be transformed into storable fuels and chemical feedstocks such as alcohols and methane gas. The actual efficiency of solar energy captured by green plants can be as much as 3–4%. The effective photosynthetic plants are maize, sorghum, and sugar cane which are the most productive. In addition, the products of photosynthesis i.e. sugar, starch, and wood (lignocellulose) are potential biomass utilised for production of biofuel. The photosynthetic plants utilise solar energy to convert CO_2 into biomass that can be used as a source of energy; such plants are called **energy crops**.

The sources of biomass should have low cost and their continuous and ample supply should be ensured to achieve optimisation of biofuel technology. There are three ways by which it is possible to have a cheap and abundant supply of biomass. These include:

a. Cultivation of energy crops.

b. Harvesting of natural vegetation.

c. Utilisation of agricultural and other organic wastes.

The resulting biomass is converted by biological or chemical processes or by a combination of both into usable form of fuel. The formation of biofuel depends on the initial biomass and the process utilised (figure 9.1).

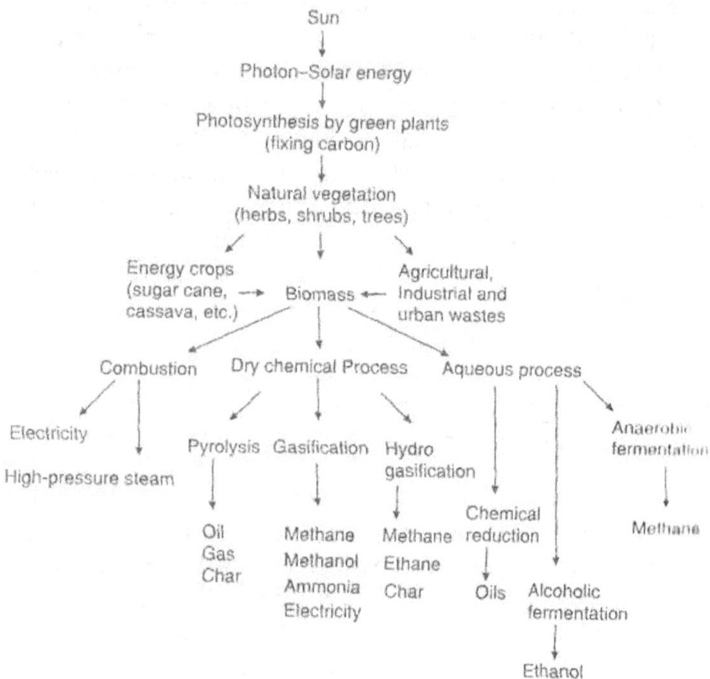

Figure 9.1   Different modes of biomass utilisation for production of biofuel

Higher yields of fixed carbon during photosynthesis in green plants is possible in well-planned plantation methods than harvesting natural vegetation or collecting agricultural or industrial wastes. Programmes for well-planned plantation are extensively practised in many countries throughout the world. In Brazil, Australia and South Africa, sugar cane and cassava are two energy crops that are being developed primarily for ethanol production, whereas in Sweden and America, lignocellulose-based programmes are being employed to grow forests for production of liquid biofuel. In India, energy crop plantation programmes have to face several problems including unseasonable rainfall, arid and drought conditions, low levels of ground water, inadequate rainfall, etc.

The processing of biomass for fuel production depends on many factors such as moisture content and chemical complexity. The biomass with high water content normally undergoes aqueous processing since there is no need for substrate drying. Ethanol is produced by hydrolysis and fermentation; methane is generated by anaerobic digestion. Chemical reduction of gaseous biomass can produce oily hydrocarbons. Biomass with less water content such as wood, straw, bagasse, waste fodder, farm weed, stubble and sawdust can be directly burnt to produce heat or to raise steam for generation of electricity; similarly it can be subjected to thermo-chemical processes such as gasification and pyrolysis to produce biofuels such as gaseous oil, char and eventually methanol and ammonia.

Truly, vast amount of low-cost, pollution-free and abundant biofuel could be obtained from the bioprocessing of biomass by different ways. Even the genetic and agronomical improvement in energy crops would contribute to reduce the production cost of biofuels.

9.3 PRODUCTION OF BIOETHANOL

Bioethanol or green petrol is produced by tiny yeast cells, *Saccharomyces cerevisiae,* using biomass which is mainly composed of sugar cane and maize. Since 1980, Brazil and USA have been increasingly using bioethanol as a substitute for petrol in vehicles. Only a small investment is required to modify the ordinary petrol engine of an automobile to make it burn ethanol efficiently. Due to governmental subsidies, ethanol is available at a little more than half the price of petrol. Vehicles can be powered even with a mixture of 20% ethanol and 80% petrol known as gasohol, and this does not require any modification of the engine. Brazil's undoubted success in pioneering the production of 'green petrol' has created worldwide interest and now many countries including India are following this trend.

Bioethanol is the product of alcoholic fermentation of sugar, starch or cellulosic materials in the presence of an enzyme zymase secreted by the yeast, *Saccharomyces cerevisiae* (figure 9.2). It is an ancient art and is often considered as one of the first microbial processes used by human beings. The overall chemical steps involved in the production of bioethanol are represented in figure 9.2.

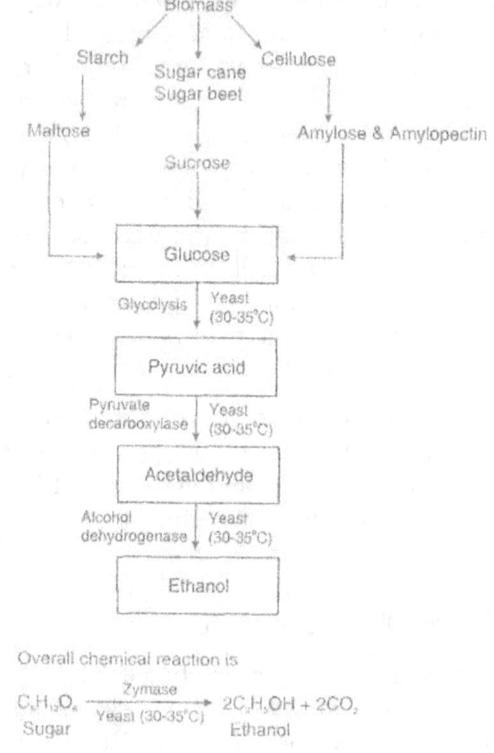

Figure 9.2   Schematic representation of chemical reactions involved in bioethanol production

9.3.1 Bioethanol Vs Petrol

Bioethanol is energy-efficient, it does not produce toxic carbon monoxide during combustion in engines and therefore is much less polluting than conventional fuels. It is cheaper to produce ethanol by fermentation processes than to produce it by chemical means (which involves hydration of ethylene). Because of these features, bioethanol is used as a good substitute for petrol. In spite of this, the engines powered by ethanol suffer from some disadvantages including the following:

1. Engines may have starting problems when atmospheric air is cool, which is due to the higher latent heat of vapourisation of ethanol.

2. Since ethanol is hydrophilic, it can absorb moisture from atmosphere, and this interferes with the combustion and may cause corrosion in storage tanks and engines.

3. Ethanol reacts with metals used in alloys of carborators of the vehicle.

4. The important reason for limiting the use of bioethanol as fuel is the high cost of downstream processing to recover ethanol that ultimately raises its production cost.

9.3.2 Biomass for Ethanol Production

Production of industrial alcohol that is used for industrial, laboratory and fuel purposes depends upon the use of biomass in the form of

1. Wood (lignocellulose) of the fast-growing trees

2. Sugar cane and sugar beet

3. Starch crops

Fermentable sugars are used as raw materials (table 9.1) after some degree of pretreatment, depending upon their chemical composition. The Brazilian National Alcohol Programme is producing over 4 billion litres of alcohol a year, mainly from sugar cane juice, while maize is the favoured raw material in the US. Both of these energy crops have the drawback that they have a relatively high value as food for human and animal consumption. A wide range of plants including plants with starchy roots, (e.g. Mandioca and potatoes) is receiving close attention as alternatives for biofuel generation. Yet another potential biomass for ethanol production is banana. National Research Centre for Banana (NRCB), Tiruchirapalli, Tamilnadu, has successfully attempted to convert the ripen epicarp of banana (which is to be a waste) into ethyl alchohol. For commercial production of bioethanol from banana waste, NRCB is planning to construct a plant at Thottiyum in Tamilnadu.

Plants use starch as an energy store and when they need to mobilise it, they take the help of the enzyme amylase that breaks down starch into sugar (glucose). Amylases are present in certain strains of bacteria. Some yeasts do possess amylases but unfortunately they manufacture little alcohol. There are two possible ways to overcome this obstacle, the first is to widen the search for microbial recruits to the biofuel industry and the second is to use genetic engineering to create yeast that can do what is required. With the advent of rDNA technology it is now possible to isolate the gene encoding amylase from the bacterial strain and introduce it into the most highly productive strains of yeasts. Use of recombinant yeast for fermenting starch leads to production of vast quantities of alcohol without any expensive pretreatment.

Similar efforts are made by genetic engineers to isolate the gene encoding cellulase from the fungal strains, *Trichoderma reesi, Coriolus hirsuitus,* and *Polyporus anceps* and introduce it into yeasts so that they can feed on cellulose present in the wood. These recombinant yeasts would pay enormous dividends.

Table 9.1   A list of potential biomass for bioethanol production

Nature of biomass	Examples
Sugar-containing	Sugar cane, sucrose, invert sugar, molasses, lactose, whey, glucose, sulphite waste, sugar beet, sweet sorghum
Starch-containing	Root and tuber crops e.g. potatoes, cereal-corn, sorghum, wheat, barely milling products, wheat flour, wheat mill feed
Cellulosic	Wood derived from plant species like *Buten, Casuarina, Leucaena, Eucalyptus, Melia,* etc. Saw dust, waste paper, straw, bagasse, forest and agricultural residues, municipal solid wastes.

9.3.3 Alternatives to Yeasts

In one of the approaches to find the alternatives to yeasts, the use of heat-tolerant bacteria, *Clostridium thermocellum* and *Thermoanaerobacter ethanolicus,* would bring four potential benefits for biotechnological production of bioethanol.

1. When the temperature of the reaction mixture is increased to obtain optimum fermentation, these bacteria would convert the raw materials (sugar, starch or cellulose) into alcohol more

rapidly than yeasts do. But the yeasts cannot withstand high temperature, they die due to the high temperature.

2. There would be no need for complex cooling systems in the fermenting vessel. In fermentation by yeasts, the fermenter must be cooled to ensure that the heat produced during fermentation does not harm the yeasts.

3. Fermentation vessels operating at relatively high temperatures are less likely to become contaminated with unwanted organisms, since not many microbes can withstand the heat.

4. As the fermentation liquid is warm, less heat needs to be provided to distil the alcohol and the cost of distillation is a significant factor in the total production cost of bioethanol.

9.3.4 Production of Ethanol by Fermentation

In a typical fermentation process to produce bioethanol, 100 g of glucose yields 45–49 g alcohol in the presence of yeast strains *Saccharomyces cerevisiae* and *S. carlsbergensis.* Both batch and continuous fermentation processes are used and often yeast cells are recycled. In continuous process of fermentation, 3–5 closed vessels are used; ethanol concentration increases from 4% in the first vessel to 10% in the final vessel; the productivity of ethanol ranges from 10 to 20 kg/ m^2/hr.

In contrast to the batch process the productivity ranges between 1.5 to 3 kg ethanol/m^2/hr. However, in batch process, the final yields are much higher and it is the most widely used method of fermentation to produce bioethanol. Generally in the substrate, ammonium sulphate or urea is added as nitrogen source and a salt of phosphoric acid as phosphate source are added. The temperature of the reaction mixture during fermentation is kept within 30–38 °C and pH is 4.5 to 5.0. An effective cooling system is necessary in large fermenter vessels.

Generally increasing levels of ethanol progressively inhibit the production process and yeast cell growth. The ethanol inhibition can be prevented by continuous removal of ethanol by the commercial process called Biostill technique in which a cell-free broth passes through an evaporation chamber for removal of ethanol and this medium is then fed back into the fermenter.

Key Concepts

- Biofuels are the fuels generated from the biomass taking the help of biological agents. They are considered as substitutes for the rapidly depleting, nonrenewable and ever-expensive fossil fuels.

- Biofuel technology deals with production of pollution free, cheap and substantial quantities of biofuels from biologically derived substrates acted upon by microorganisms.

- Energy crops are photosynthetically derived plants that utilise solar energy to convert CO_2 into biomass, which can be utilised as a source of energy.

- Biomass can be converted by biological or chemical processes or by a combination of both into useful biofuel such as bioethanol, methane gas or hydrogen.

- Brazil pioneers in the production of bioethanol or green petrol.

- Bioethanol is a good substitute for petrol, and it can also be mixed with petrol to form gasohol which is also an alternative to petrol.

- Ethanol has much application in laboratory and industries. But the production of ethanol by yeast is too limited, that it cannot be implemented for large-scale industrial production until recombinant yeast having genes, encoding for cellulase and amylase are produced.

The recovery of ethanol is based on distillation. The broth is distilled in a beer column where the alcohol present in the broth is vapourised by steam. This results in the formation of 85% v/v ethanol. The next step is to rectify the ethanol to 96%. It is then dehydrated to 100% alcohol by addition of a small amount of benzene. The downstream processing for recovery of absolute alcohol requires a considerable amount of energy that ultimately increases its cost of production and it is due to this fact that bioethanol has limitation as biofuel for automobiles. The flow diagram for bioethanol production from various biomasses is shown in figure 9.3.

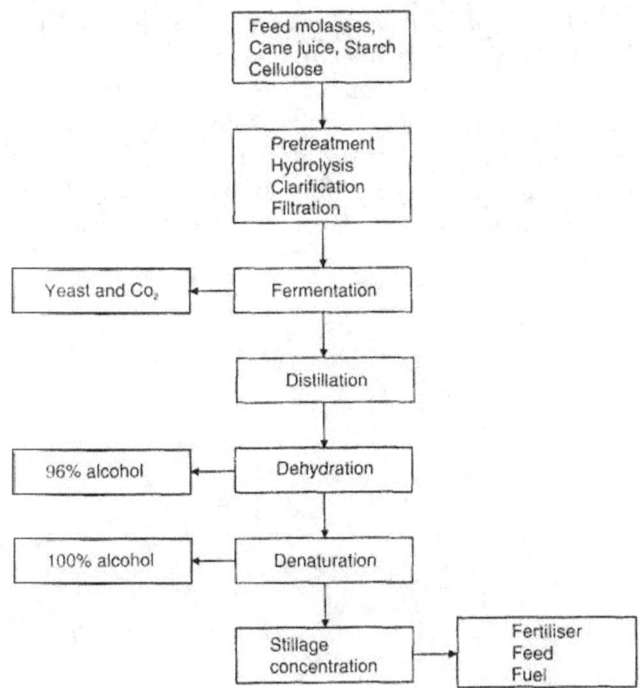

Figure 9.3   Flow diagram for bioethanol production

9.4 PRODUCTION OF METHANE FROM BIOMASS

Methane gas is a product of anaerobic fermentation of biomass. It is also called biogas or gobar gas since cattle dung is largely used in its production. The process of production of biogas is also called methanisation. In addition to cow dung, human excreta (upto 3% of slurry) and kitchen waste can also be used as raw material. Methane gas can be used for generation of mechanical, electrical and heat energy, and is now extensively used as a fuel source for domestic and industrial purposes.

The microbiology of methane production is complex, involving a mixture of anaerobic microorganisms (figure 9.4).

Biogas is produced with the help of the following three groups of bacteria.

1. *Hydrolytic or decomposer bacteria*   It is the first group of
   bacteria such as *Clostridium, E.coli,* and *Bacillus.* They produce
   enzymes that convert complex, insoluble organic substances
   (polymers) into simple soluble substances (monomers) by
   hydrolysis. Proteins, fats, cellulose present in raw material are
   converted into simple amino acids, fatty acid, sugar etc.
   However, lignins are not decomposed and remain as residue.

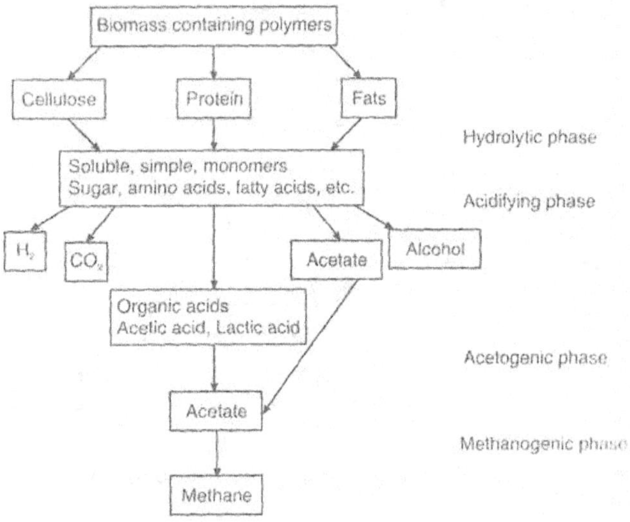

Figure 9.4   Sequential anaerobic breakdown of biomass for production of biogas

2. *Acetogenic bacteria*   It is the second group of bacteria that
   include acid-forming bacteria like *Sytrophomonas,*
   Disulfovibrio, etc. They convert compounds formed in the first
   step (monomers) into organic acids like acetic acid, lactic acid,
   etc.

3. *Methanogenic bacteria*   This is a third group of bacteria which
   convert organic acid (acetic acid) to form methane and CO_2.
   e.g. *Methanococcus, Methanothrix.*

The biogas thus produced mainly consists of 50–70% methane,
30–40% CO_2, and traces of H_2 and N_2.

The device constructed for the production of biogas is known as
biogas plant. In India most popular designs have been developed by

Khadi and Village Industries Commission (KVIC), New Delhi, (figure 9.5). All types of biogas plants have the following basic parts:

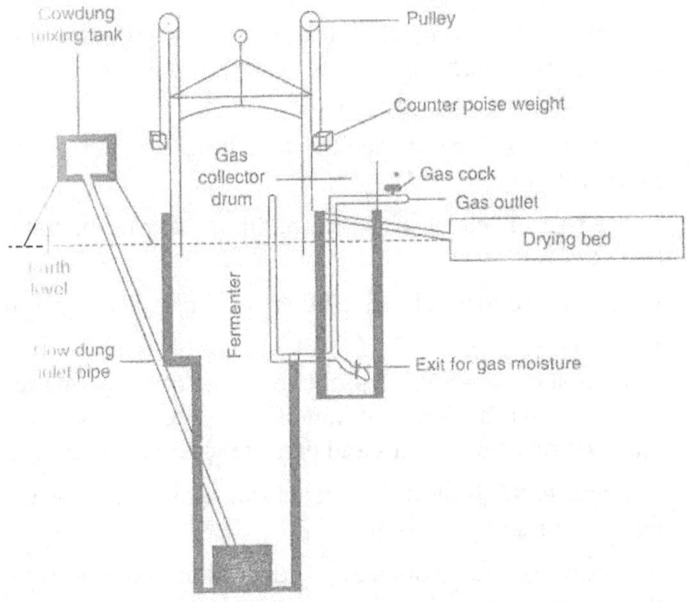

Figure 9.5   Design of gobar gas plant developed by I.A.R.I.

The main part is called gas digester where the gas is produced; it is an air-tight underground cylindrical tank which is further connected via pipes with two small tanks present at ground level on either sides of the gas digester. One is the inlet tank used to add gobar slurry into the digester where anaerobic bacteria sequentially convert the raw material into biogas. The gas is collected into a biogas holder which is an inverted gas collector drum made up of a steel sheet. From this the gas is then removed as per requirement. The digested slurry comes out through the outlet pipe into the outlet tank. This residue is rich in nutrients and is used as fertiliser or soil conditioner. For good production of gas a pH of about 6.8 and a temperature of 38 °C are required. It takes 15 to 30 days for the formation of the gas.

9.4.1 Advantages of Biogas

1. It is a simple low-cost technology providing alternative source of energy in rural areas.

2. It can substitute wood, dung cake, electricity, L.P.G., etc.

3. It is a clean biofuel that burns without smoke and hence is pollution-free.

4. The recovery of product (methane gas) is spontaneous as it automatically separates from the slurry. It does not require any costly downstream processing.

5. The technology is eco-friendly since it utilises biomass which may cause organic pollution.

6. There is no need of asceptic conditions during operation of biogas plant.

7. Biogas is used for cooking, lighting, heating boilers, firing brick, etc.

8. It improves sanitation by proper disposal of waste matter and also prevents incidence of waterborne diseases as it involves inactivation of pathogens and parasites by anaerobic digestion.

9. The nutrient-rich residue released from gas digester is used as fertiliser or soil conditioner.

10. It is a storage source of energy with minimum risk of explosion than pure methane.

However, there are still many inherent problems associated with the production of biogas. Some of them are as follows:

1. The plant does not work at low temperatures particularly during winter.

2. The cost of collection of organic matter for the purpose of methanogenesis on a large industrial scale is too expensive and the rate of biogas production is inconsistent and low, therefore the process is uneconomical.

3. The presence of lignin in the biomass of most agricultural and urban wastes creates another problem since it is not easily digested by anaerobic processes; and physical and chemical pretreatment places a considerable energy-and-cost burden on the overall process.

4. Microbial production of methane is more expensive than natural gas.

5. Biogas cannot be used in automobiles as it contains some gases as impurities that are corrosive to metal parts of engines and conversion of this biofuel to liquid state is difficult and expensive.

9.5 PRODUCTION OF BIO-HYDROGEN

The biotechnological research into hydrogen production offers a potential fuel that has a high energy-to-mass ratio and that can be produced from a power supply that will last for millions of years (the Sun) and from a raw material that covers three-fifths of this planet (water). In addition, hydrogen, forms water, when burnt thereby causing no pollution and thus renewing the raw material.

The public view regarding the use of hydrogen as fuel is pessimistic because of the explosion of the Hinderburg airship (powered by hydrogen) that killed hundreds of air passengers. But the fact is that we use more dangerous fuel than hydrogen in our homes. It is the cost rather than the safety that prevented hydrogen from becoming a common fuel. It has been estimated that at least 20 to 25 years of research by biotechnologists is needed to change the present scenario.

Production of hydrogen can be possible by

i. Biological agents like bacteria and algae

ii. Photolysis of water

iii. Gasification of biomass

Hydrogen produced with the help of a biological agent, is called bio-hydrogen. There are two ways to achieve this:

a. Fermentation of substrate by anaerobic bacteria.

b. Biophotolysis of water.

An additional potential approach is developed to generate bio-hydrogen by *in vitro* system by coupling of chloroplast and the enzyme hydrogenase (isolated from algae or bacteria) in the presence of sunlight.

9.5.1 Anaerobic Fermentation Producing Hydrogen

Biomass comprising cellulose or starch is hydrolysed enzymatically or chemically to form the simple sugar glucose, which is then oxidised by anaerobic bacteria like *Clostridium butyricum* by reducing NAD^+ to NADH (figure 9.6a). It is necessary to remove NADH from the cell environment and regenerate NAD^+ to continue the oxidation of glucose. The electrons from NADH are transferred to H^+ ions to produce H_2 gas, thereby regenerating NAD^+. The reaction is catalysed by the enzyme hydrogenase generated by anaerobic bacterial strains that are supplied with sugar (figure 9.6b).

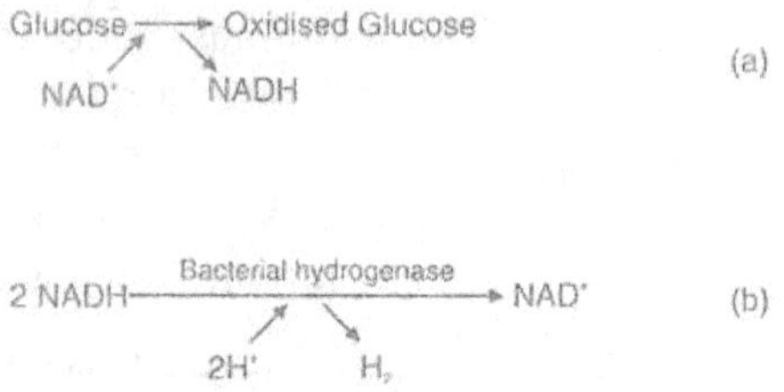

Figure 9.6 Reaction involved in H_2 production during anaerobic fermentation of glucose

9.5.2 Photosynthetic Algae Producing Hydrogen

Chlorella pyrenoidosa is the algal strain used in the experiment to produce hydrogen gas in the presence of sunlight. When the algal strain is illuminated, it can assemble two water molecules and two carbon dioxide molecules to form a compound called glycolate which is then fed to an immobilised enzyme, glycolic oxidase (obtained from plants) to convert glycolate into formate.

$$2CO_2 + 2H_2O \xrightarrow[\text{Photosynthetic algae}]{\text{Light}} CH_2OHCOOH + 1.5\ O_2$$

$$CH_2OHCOOH + O_2 \xrightarrow{\text{Glycolic oxidase}} CHOCOOH + H_2O_2$$

$$CHOCOOH + H_2O_2 \xrightarrow{\text{Non-enzymatic}} HCOOH + CO_2 + H_2O$$

$$HCOOH \xrightarrow{\text{Formic hydrogenase}} H_2 + CO_2$$

In all $$2H_2O \longrightarrow 2H_2 + O_2$$

Figure 9.7 Stages in production of hydrogen by photosynthetic algae

In the final stage of the process, formate is supplied to bacteria that have been fixed onto glass beads. They consume the formate and release hydrogen gas and carbon dioxide, thus separating the H_2 and O_2 that were originally combined in the water molecule (figure 9.7).

9.5.3 *In vitro* Production of Hydrogen

An *in vitro* system was developed by the combination of stable chloroplasts and hydrogenase-producing bacteria. Chloroplasts extracted from plants trap solar energy with the help of chlorophyll molecules. The energy is used to split water molecule (Photolysis of H_2O) into two H^+ ions and an oxygen atom, and simultaneously two electrons (e^-) are released. The H^+ ions diffuse from the chloroplasts and the electrons are transported by a series of molecules known as electron carriers. Hydrogenase enzymes obtained from bacteria are used to form H_2 molecule from two H^+ ions and two electrons. The hydrogen gas thus formed, bubbles out of the solution and is collected (figure 9.8). The major problem of the system lies in the sensitivity of hydrogen to the O_2 produced during water splitting.

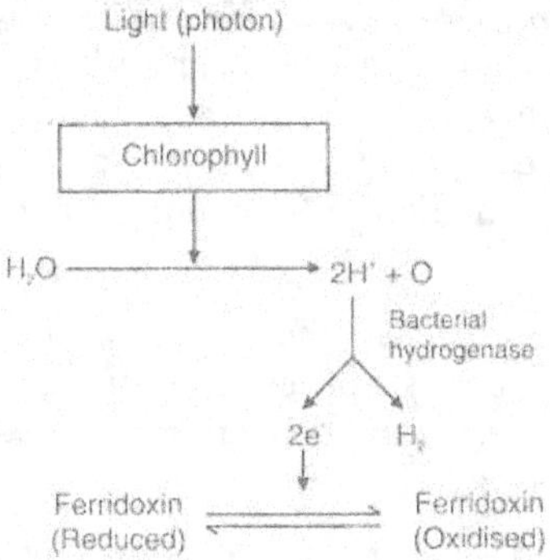

Figure 9.8   *In vitro* system comprising chloroplasts and hydrogenase in the presence of light to produce H_2

In addition, scientists at the National Environmental Engineering Research Institute (NEERI) have identified the photosynthetic bacteria that use solar energy to generate hydrogen from waste water from ice cream and butter factories. With this technology, H_2 production is coupled with removal of pollutants. The gas can be used in edible oil industry for hydrogenation of vegetable and animal fat. The gas has immense potential for use as a chemical feedstock in production of NH_3, methanol or other chemicals.

Although it is possible to generate hydrogen by the above-mentioned routes, the production rate is too small to make microbial genesis of hydrogen economical. The efficiency of hydrogen production by anaerobic fermentation is much less than that of methane production by the same method. The purely biological materials are just too fragile for industrial production of hydrogen. However, further research in this field may well alter these considerations.

In addition to the above-mentioned biofuels, biodiesel is also gaining public interest. Several plant species belonging to the family Euphorbiaceac (e.g. *Euphorbia lathyris*) which produce latex enriched with hydrocarbons are used to manufacture biodiesel. In Lucknow, a governmental organisation has successfully produced biodiesel, from *Jatropha gossypifolia* and *Jatropha glandulifera* that is used for running diesel engines of the Northern railway.

Key Concepts

- Production of methane gas or gobar gas is a low-cost, simple and eco-friendly technology which involves the role of hydrolytic, acetogenic and methanogenic bacteria in anaerobic fermentation of cow dung, human excreta and kitchen waste.

- The device constructed for the production of biogas is known as biogas plant consists of a gas digester and a gas holder where gas is collected and then utilised for cooking, lighting, heating boilers, etc.

- The process does not require aseptic conditions; the recovery of product is spontaneous and the process releases nutrient-rich residue that is used as fertiliser. But the process, however, has some inherent problems.

- Production of hydrogen is possible via fermentation of substrate using anaerobic bacteria and photosynthetic algae.

- *In vitro* production of hydrogen is also possible using isolated chloroplast and hydrogenase in the presence of sunlight.

- Since the efficiency of production of biofuels using fragile biological agents is very low further research in this field is very much essential.

Review Questions

1. Define biofuel technology. Explain the different sources of biomass with their mode of utilisation.

2. What are biofuels? Explain the method of production of bioethanol.

3. Define biogas. Explain the role of bacteria in anaerobic fermentation of biomass for the production of methane gas.

4. What is gobar gas? Explain the design of biogas plant with the functions of its components.

5. How is biogas advantageous compared to other conventional fuel? Enlist its disadvantages.

6. Explain the role of rDNA technology in the modification of yeast to achieve commercial production of ethanol from low-cost substrates.

7. Describe the various processes of production of hydrogen.

8. Write short notes on:

   a. Heat-stable bacteria in ethanol production

   b. Methanogenesis

   c. Biomass for ethanol production

   d. *In vitro* production of H_2

10

ENVIRONMENTAL BIOTECHNOLOGY

10.1 INTRODUCTION

Man is living in a world that is urbanising and industrialising at an enormous pace. In succeeding in his objective to uplift his standard of living, he has also deteriorated the environment. The environment is the sum of all the factors including both biotic (living organisms) as well as abiotic (air, water, temperature, etc.) factors. Now the public concern is mounting over the untoward changes that have occurred in the environment and it is the need of time to improve the environment for the next generation. In developed countries, this objective is achieved because of the major environmental legislation directed toward liquid, solid and hazardous waste, while in developing countries, the situation is less encouraging due to limited financial support to construct water treatment and waste treatment facilities. In addition, there is lack of official regulation and control systems, lack of administration bodies to control waste and little obligation for existing and emerging industries to dispose of waste properly.

Man borrowed life-support systems comprising oxygen, water, energy, raw material, nutrients and living place from the environment and in return due to various activities like domestic, agricultural, manufacture, recreation, transport, bioterrorism, nuclear tests, warfare, accidents, etc. generated waste and added pollutants to contaminate air, water and soil—the three significant abiotic components of the environment. In the final assessment, wastes generated from various human activities represent the end of the technical and economic life of a product. Costs for properly dealing with wastes are progressively increasing and much attention is now paid to efficient and effective waste management. Environmental biotechnology involves biotechnological approaches to the management of the environment and socio-economic problems. Many successful biotechnological methods have been developed for water, gas, soil and solid waste treatments. All such methods make use of metabolic (degradative and anabolic) activities of microorganisms— once again microbes are playing an indispensable role in our ecosystem in cleaning up our environment.

10.2 SOURCES OF WASTES AND POLLUTANTS

Waste is a man-made worthless matter in the form of any product, by-product or residue while pollutant is a waste product that damages the environment. Generally all pollutants are wastes but all wastes are not pollutants. Wastes can be biological, chemical, geo-chemical or physical agents that are released intentionally or inadvertently by various man-made activities into the environment in such concentrations that they may have adverse, harmful or unpleasant effects. The various human activities which act as sources of wastes and pollutants are listed below:

1. Manufacturing   The nature of raw material used in the manufacture of a desired product, and the design and operation involved in the process are influencing factors in the generation of a wide variety of wastes. Industries like food, agriculture, dairy, oil refinery, paint, breweries, fertilisers, detergent, leather and wool, medical and pharmaceutical, textile, metal processing, mining, plastic, etc. generate a variety of refuse and by-products. Some of them are biodegradable while some are persistent and obnoxious. The pollutants released from the industries include inorganic and organic compounds, oils, acids, phenolics, volatile hydrocarbons, aerosols, heavy metals, cyanides and different oxides of sulphur and nitrogen, pesticides, feed residues, etc.

2. Transport   From a global point of view, the transport sector is the biggest contributor so far to environmental pollution. The people from some of the metropolis cities like New York, Delhi and Tokyo are the worst affected by the ill effects of pollutants generated by transport. The scrap vehicles, used-up oil, grease, etc. are some of the wastes, and atmospheric pollutants like CO, volatile hydrocarbons, soot, lead and oxides of sulphur and nitrogen are adversely deteriorating the environment whether it may be air, water or soil. Pollution caused by the transport sector poses the greatest challenge to biotechnologists. To some extent biotechnologists have a solution to this ever-increasing problem by way of biofuels that are less polluting than conventional fossil fuels.

3. Domestic and house-building activities   Human faeces and urine are the chief domestic refuse which become a part of sewage released without proper pretreatment into or in the vicinity of water bodies in most of the places in India as well as other developing countries. The domestic garbage comprising kitchen wastes, bottles, tins, plastic, paper waste, etc. are collected and dumped into large pits (landfill) or on the open barren land around the cities creating further problems of air, water and soil pollution. Some of the pollutants percolate and contaminate the ground water. In many parts of the world there is increasing evidence that underground water sources are demonstrating dangerous levels of contamination.

The demolition of old buildings and construction of new ones generates wastes in the form of wood, asbestos, metals, glass, plastic, fly ash, etc. Raising commercial buildings and factories also contributes to environmental pollution in one or other way. The intensity of pollution caused by the demolition of commercial buildings was experienced by the Americans when terrorists attacked the World Trade Center situated at Manhattan in New York on September 11, 2001.

Key Concepts

- Urbanisation and industrialisation are two important factors responsible for increase in environmental pollution. In developing countries the problem is more serious due to lack of strict regulations, financial support and obligations for existing and emerging industrial plants to dispose of waste properly.

- Industries manufacturing different products generate variety of refuse and by-products.

- Transport is another sector that causes environmental pollution.

- Demolition of old buildings and houses and subsequent building construction activities also add considerable pollutants to our surroundings.

- Energy production from fossil fuels produces wastes, which ultimately cause detrimental changes in the ecosystem.

- Medical activities in the hospitals produce dangerous biomedical wastes that create serious problems in disposal. The physical, chemical and biological pollutants produce variety of hazards in the living components as well as in the abiotic factors of the ecosystem.

4. Energy production   The entire world depends on fossil fuels for the production of energy, which leads to release of huge quantity of CO_2, CO, different oxides of sulphur and nitrogen, and heat into the environment. When radioactive elements are involved in energy production (usually at nuclear power stations), it requires a very safe method for disposal of radioactive wastes. Photosynthesis enables recycling of CO_2, but continuously increasing CO_2 concentration may cause global warming which has its own drawbacks (green house effect). In addition, acid rain is a consequence of SO_2 and NO_2 in the atmosphere, which ultimately affects our ecosystem. Energy plants utilising coal as the chief ingredient produce bulk of the fly ash; its disposal creates severe problems as it contaminates soil and water bodies.

5. Medical activities Hospitals generate biomedical waste in the form of syringes, needles, vials, polythene bags, saline bottles, and even bodily refuse such as patients' blood, pus, sputum, plasters, rejected tissues and body parts, etc. These create serious problems in disposal. In some cities there are kilns to burn all such obnoxious and hazardous wastes but such practices cannot wipe out the problem of environmental pollution since their burning in kilns release other air pollutants which create further health hazards.

10.3 HAZARDS PRODUCED DUE TO WASTES AND POLLUTANTS

The World Health Organization (WHO) considers wastes and pollutants causing both short-term hazards and long-term environmental hazards. Short-term hazards include acute toxicity by ingestion, inhalation or skin absorption, corrosivity or other skin or eye contact hazards or risk of fire or explosion. Long-term environmental hazards include chronic toxicity upon repeated exposure, carcinogenicity, resistance to detoxification processes such as biodegradation, the potential to pollute underground or surface waters or offensive smells at aesthetically objectionable proportion.

The various human activities responsible for generation of different kinds of wastes and pollutants can be broadly categorised into 3 groups causing various hazards.

10.3.1 Hazards Produced Due to Biological Agents Present in Wastes

Biological entities present in biological wastes include living organisms and viruses (natural and genetically engineered) that infect humans, animals and plants and cause diseases. They may also cause acute toxicity and allergy. This may occur due to contamination of food, water or animal feed by the waste or may be mediated by vectors like mosquitoes, flies, fleas, etc. (table 10.1).

Table 10.1   Communicable diseases of man and animals

Causative agent	Disease
1. Viruses	Epidemic jaundice (Infective hepatitis), pseudorabies, bluetongue, bird flu
2. Prion	Mad cow disease
3. Bacteria	Cholera, typhoid, diarrhoea, bacillary dysentery, gastroenteritis, foot and mouth disease, brucellosis
4. Fungi	Candidiasis
5. Entamoeba	Amoebiasis
6. Plasmodium	Malaria
7. Trypanosoma	Sleeping sickness
8. Helminths	Filariasis, trichinellasis, ascariasis

In addition the microbes can be leaked out or may escape from the process directly into the environment. Medical facilities provided through health centres, proper sanitation and waste disposal system, vaccination and immunisation programmes, etc. can decrease ill effects.

10.3.2 Hazards from Chemicals Present in Wastes

Chemicals like halogenated aliphatics, polynuclear aromatics, heterocyclic polar non-halogenated nitrocompounds, hydrocarbons, chlorofluorocarbons (CFCs), cyanides (sodium cyanide, hydrocyanic acid, methyl isocyanides, etc.), phenolics, trichlorethylene, benzene, styrenes, pentachlorophenols (PCPs), polychlorinated biphenols (PCBs), heavy metals, organochloro pesticides, H_2S, NH_3, chloride, fluoride, etc. are some of the chemical pollutants present in effluents, solid wastes or gases; they may be toxic or nontoxic, and biodegradable or nonbiodegradable. Microorganisms can act upon few chemical compounds to convert them into useful biological and/or chemical resources.

Xenobiotics are synthetic compounds not formed by natural biosynthetic processes and in many cases, can be recalcitrant. Xenobiotic compounds like organochloro pesticides (BHC, DDT, etc.), PCPs, PCBs, CFCs, synthetic fibres, plastics, polyethylene detergents, etc. are foreign substances in our ecosystem that persist in nature and may often have toxic effects. These compounds are not degraded in nature even when conditions appear to be adequate for microbial growth, and hence such compounds are called recalcitrant.

Biomagnification is the phenomenon of progressive increase in concentration of nondegradable synthetic compounds with the rise in trophic level in food chain. For example, DDT (Dichloro diphenyl trichloroethane) absorbed by organisms of lower trophic level (plant and microbes) may be eaten by the organisms of higher trophic level including fish. Here the concentration of xenobiotic compound increases in body of fish because it consumes more food from lower level. When bird eats the fish the concentration of toxic compound further increases in bird. This is called biomagnification. As a result, DDT is progressively accumulated in the body of every organism of higher trophic levels causing toxic effects. Sea birds accumulate DDT in their bodies as a consequence of bioaccumulation and due to this they lay eggs with fragile shells and mass destruction of eggs results in reduction of population. Being one of the members of the higher trophic level in the food chain, man too suffers the impacts of biomagnification. Nowadays, it is a general saying that even the mother's milk is polluted with DDT.

Chemical pollutants contaminate water, soil and air causing serious health hazards like digestive disorders, respiratory impairments, vision problems, skin lesions and even cancer, thus providing an immediate threat to man and the ecosystem.

10.3.3 Hazards Produced Due to Physical Pollutants

Physical pollutants are dust, humidity, radiation, explosive elements, sound, etc. Radiation acts at the molecular level in the cell causing hazardous effects on biological systems; its impact are memorable. The atomic bomb explosion at Hiroshima and Nagasaki during the Second World War is a classsical example for understanding the impact of radiation hazards. In addition to mental stress, the risk of fire, explosion due to dangerous explosive wastes and hearing impairments in human beings caused by sound pollution are other hazardous effects of physical pollutants.

10.4 MICROBIAL DEGRADATION OF XENOBIOTICS

Xenobiotics are man-made chemical compounds that persist in nature. Microorganisms can degrade most of the xenobiotics while the compounds that resist biodegradation and persist in the environment are called recalcitrant. Many recalcitrants are toxic to bacteria, lower

Table 10.2   Microorganisms having biodegradation potential

Microbe	Toxic pollutant
Acinetobacter sp.	Chlorobiphenyl
Alcaligenes sp.	Chlorophenols, PCPs, PCBs
Arthrobacter sp.	Hydrocarbons, benzene, polycyclic aromatics
Aspergillus sp.	Chlorobenzoate-herbicides and chemical pollutants in tannery effluent
Azotobacter sp.	Aromatics
Candida sp.	PCBs, formaldehyde
Flavobacterium sp.	Organophosphates
Fusarium sp.	Propanil, cyanide, phenol
Methanococcus sp.	Haloethanes like bromo and chloroethanes, trichloroethylene, etc.
Mycobacterium sp.	Benzene, aromatics, cycloparaffins
Nocardia sp.	Cyclohexane
Phanerochaete sp.	DDT, lindane, PCB, PCP, coal tars and heavy fuels
Pseudomonas sp.	Hydrocarbons, PCBs, alkyl benzoates polycyclic aromatics, organophosphates (like malathion, parathione) benzene phenolics, naphthalene, anthracene, cyanides
Trichoderma sp.	Malathion
Xanthomonas sp.	Hydrocarbons, polycyclic hydrocarbons

eukaryotes and even man, and some of them have been shown to be carcinogenic. The xenobiotic compounds like DDT and PCBs are recalcitrant and lipophilic and as a result they show biomagnification as described in Section 10.3.

The rate of natural biodegradation is often slow and mainly depends upon pH, temperature, availability of oxygen, type and number of microbes involved, complexity and concentration of pollutants, toxic intermediates produced in the process, etc. For optimising the rate of biodegradation, a mixed consortia of microbes is employed and further nutrients and cofactors are added to allow better adaptation and more tolerance in microbes. A wide variety of inducible enzymes are produced when a community of different species is employed for the degradation of a mixture of pollutants. Since enzymes are the products of the gene, certain strains of bacteria are specially designed by gene manipulation technique to do a specific function i.e. degradation of a specific xenobiotic. The bacterial strains like *Pseudomonas, Micrococcus, Azotobacter,* methanogens, some yeast and few fungal strains have relatively good ability to detoxify many of the xenobiotics. Table 10.2 represents a list of microorganisms having biodegradation potential for specific toxic chemicals.

10.5 MANAGEMENT OF HAZARDOUS WASTE

Anything that is discarded is a waste generated by human commercial industrial activities. As stated earlier waste may be solid, liquid or gas and may occur in dispersed, diluted or concentrated form. In addition, waste may be degraded by microbes, may persist in nature or may be a mixture of both. There are various options for the management of waste that can be categorised as follows:

1. Solid waste can be managed by

    a. Landfill technique

    b. Thermal technique—burning or incineration

    c. Composting with aerobic microbes

2. Liquid waste can be treated by

    a. Aerobic digestion

    b. Anaerobic digestion

3. Gaseous pollutants can be managed by biofilters.

10.5.1 Landfill Technique

This technique is used to treat solid wastes that are generated by modern urban society. The solid wastes contain decomposable solid organic material such as paper, food wastes, sewage wastes, glass, plastic and wastes from large-scale poultry and farmhouses. In large cities, the disposal of huge solid waste is problematic and can be solved to some extent by low-cost anaerobic landfill technology. In this process, a natural or man-made pit at a low value site is filled with everyday waste and covered by a layer of soil. The complete filling of such pits in unused areas can take months or years depending on the size of the pit and the input of wastes.

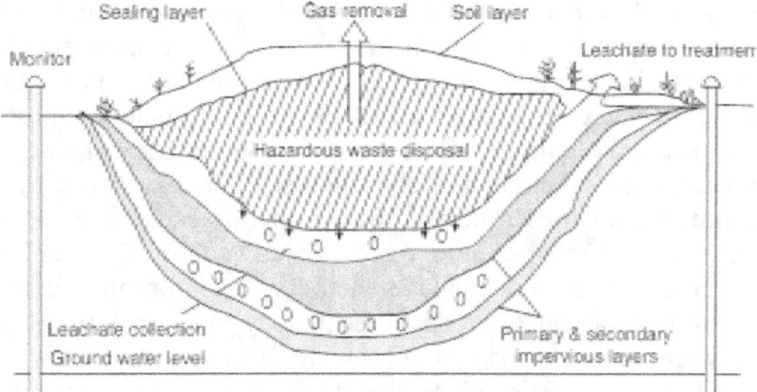

Figure 10.1   Well-sealed landfill site

Improperly managed waste produce bad odour and the site becomes unhygienic. To avoid this the waste can be pretreated before its deposition. The sorting of wastes, mechanical pulverisation or even burning are the processes to be done as a part of pretreatment. Properly constructed and sealed landfill site (figure 10.1) can be used to produce methane gas that can be collected and utilised for various commercial uses; improperly treated and unsealed landfill sites produce foul smell and stimulate vector reproduction and the generated gas may leak out and can even cause fire. In addition, improperly operated landfill sites may produce toxic pollutants and products due to anaerobic decomposition of waste that percolates from the site into underground water bodies. To prevent this, efforts should be made to put strong, impermeable liners to avoid leachates damaging the surrounding land and water bodies. A landscaped garden can be developed on the topsoil of landfill as a part of reclamation of the site.

10.5.2 Composting of the Solid Wastes

Composting is the process by which solid organic wastes are converted by aerobic microbes into a stable, humus-like material that can be recycled to the environment. Aerobic microorganisms carry out fermentation of solid organic wastes generated due to domestic and farmyard activities. The final product formed is mostly used for soil improvement. In a more specialised way, the specific organic raw substrates like straw, paddy, animal manure, etc. are utilised to form the final product that becomes the substrate for commercial cultivation of edible mushrooms. Composting is carried out using starting materials arranged in aerated piles or covered tunnels or in rotating bioreactors where temperature and moisture level should be adjusted to optimise maximum growth and reproduction of microorganisms. Composting is a simple, natural and low-cost technology (as compared to landfill and incineration), free from toxic emission. This economical and safe technology is employed by most of the farmers and authorities of municipalities in India. In European countries composting is a widely accepted technology to dispose municipal solid waste and to recycle them back into the ecosystem. Germany has so many composting plants that it is popularly known as 'green' nation.

10.5.3 Waste Water Treatment

The increasing population is proportionately augmenting a wider range of waste products, many of which cause hazardous effects on the biotic components of the environment, at the same time deteriorating the quality of abiotic factors including air, water, soil, etc. Waste water treatment involves aerobic or anaerobic microbes, and the process requires large sewage treatment plants or bioreactors (figure 10.2).

The biological disposal of organic wastes is carried out in different ways in different parts of the world. A widely accepted and practised sewage treatment process is complex but it has the highest success rate. The system consists of three stages of primary and secondary treatment followed by microbial digestion. The tertiary stage is employed for removal of specific material by chemical precipitation method. The primary treatment removes suspended particles and soluble compounds from the waste water, leaving dissolved organic materials in waste water collectively called settled sewage that is to be digested or oxidised by microbes in highly aerated, open bioreactors.

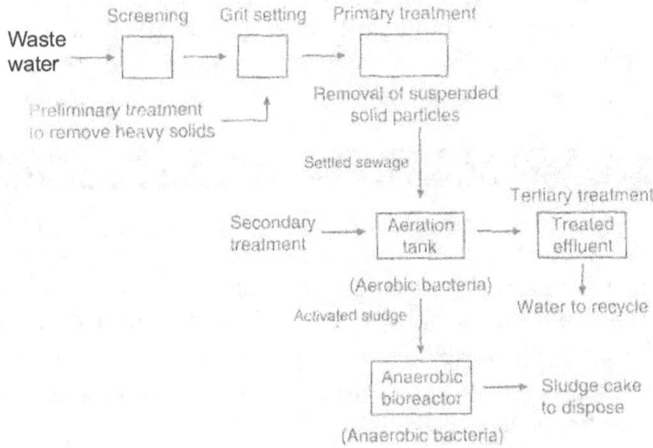

Figure 10.2 Diagrammatic representation of various stages in wastewater treatment

This is the secondary stage, which requires considerable amount of energy to operate the mechanical aerators so that there is active mixing of microorganisms, substrate and air. In this stage, aerobic microbes multiply and a biomass of sludge is formed which can either be removed and dumped in the sea or landfill or processed further in an anaerobic digester. In an anaerobic bioreactor, the volume of solids, the smell and the number of pathogenic microbes are reduced.

For treating the diluted organic liquid wastes, a percolating or trickling filter bioreactor is used. In this method the waste water flows over the surface of stones, gravels, plastic sheets, etc. on which microorganisms are present as a layer (biofilm) removing organic material from the waste water for their growth. Such techniques are widely employed in water purification systems all over the world.

10.5.4 Biofilters for Air Pollutants

There are varieties of industries, farmhouse activities, fermentation plants and poultry houses that produce waste gases with offensive odour. The major components of air pollution are sulphur oxides, nitrogen oxides, carbon monoxide, hydrocarbons, particulate matter, etc. which are responsible for deterioration of the environment, leading to serious health hazards. Biofilters provide one of the best purification systems for biological waste gas. The Federal Republic of Germany and Netherlands are leading in the field of biofiltration for deodourisation of waste gases.

Key Concepts

- The adverse effects of different types of pollutants on humans and animals include increase in the population of insect vectors, spread of diseases and disorders through contaminated food and water and chronic toxicity and carcinogenicity, etc.

- In addition, radiation impact and risk of fire or explosion are also the hazardous effects of pollutants.

- Xenobiotic compounds are man-made chemical compounds, which remain in the environment for a long time.

- The xenobiotics that resist biodegradation are called recalcitrant; many of them are toxic to bacteria, eukaryotes and man.

- These compounds show biomagnification and bioaccumulation.

- A variety of microbes like bacteria, yeast and fungal strains have capabilities to degrade the xenobiotic compounds.

- Environmental biotechnology encompasses different methods for management of hazardous wastes.

- Landfill and composting for solid waste disposal, aerobic and anaerobic digestion for waste water treatment and use of biofilters to wipe out air pollutants are some of the widely employed techniques to protect the environment from detrimental effect of different forms of pollutants.

Biofilter is a device of either a solid support or a two-phase (gas/liquid) system with suitable microorganisms to convert waste gas with offensive odour into non-hazardous compounds. In solid support system peat, soil, compost, heather, bark, etc. is used for growth of microorganisms. The waste gas is passed through the solid support layer where the gas pollutants are broken down into odourless compounds by the action of microbes present in the solid support.

In the two-phase (gas/liquid) biofilters, instead of solid support, a membrane is placed to separate the liquid phase from the gaseous phase; microorganisms are located on the membrane at the side of the liquid phase. When waste gas diffuses from the gaseous phase to the liquid phase through the porous membrane, they are acted upon by microorganisms.

Microbes employed in biofilters are mesophilic. The requisite conditions of temperature 15–40 °C, moisture 40–60% and gas contact time 1–30 seconds are maintained. Increase in humidity is the limiting factor which stops the bioconversion.

10.6 BIOREMEDIATION

Bioremediation comprises biological methods to clean up contaminated soil, ground water, seas, oceans and other water bodies. It is also called biorestoration, bioreclamation or biotreatment, and involves the role of microorganisms to carry out metabolic activities to detoxify the pollutants mainly in the form of oil-derived compounds and other toxic chemicals. The accidental spilling or leakage of oil from shipping tankers contributes in a major way to environmental pollution. In addition, industrial effluents, sewage and river outflows with increased concentration of pollutants are added to seas almost every day. It is estimated that more than 2 million tonnes of oil enter the sea each year, and it is considered that only about 18% of the total is coming from oil refineries, off-shore operations, and tanker activities. The oil spillage produces catastrophic and immediate adverse effects on birds and other aquatic life. The threatening impacts of major oil spillage were witnessed during Iran–Iraq war and during the attack made by Iraq on Kuwait, the world's largest oil producing country.

A wide range of activities associated with our modern industrialised societies produce different pollutants and contaminated soil and water bodies. Many xenobiotic compounds can recalcitrate and in many cases, a small concentration of pollutant can be subjected to biomagnification. Environmental biotechnology has to play a role in degradation and detoxification of hazardous pollutants with the help of biological agents mostly microorganisms. Bioremediation is a part and parcel of environmental biotechnology that involves the remedial activities of microbes for the treatment of environmental pollutants.

The identification of a contaminated site, the assessment of the nature and degree of the hazard and the selection of suitable remedial action are essential approaches to deal with the situation. The principle of bioremediation is simple and includes optimisation of the environmental conditions for the microbes to clean up the environment rapidly and completely. Bioremediation is preferably carried out using indigenous microflora but their rate of conversion of hazardous pollutants into useful biodegradable chemicals is very slow. Therefore it is customary to apply a consortia of microbes (inoculum) having rapid degradative activities. This process is called bioaugmentation that plays a significant role in speeding up the rate of biodegradation of a toxic pollutant.

There are two methods by which bioremediation of polluted soil, sediment or ground waters can be carried out.

1. Rapid growth of microorganisms *in situ* can be achieved by addition of nutrients. When the indigenous microbial population is exposed to a specific pollutant for prolonged period due to limited metabolic activity the rate of degradation is slow and can be accelerated when essential growth-stimulating nutrients like nitrogen and phosphorus are added. This ultimately promotes the growth of microbes and subsequently the rate of biodegradation is increased. This method was successfully employed to clean up the oil spillage from the oil tanker Exxon Valder at Prince William Sound and the Gulf of Alaska during 1989–90. Use of warm-water spray on rocky shore, and skimming and vacuuming the oil that has gone back into the water are the normal methods to clear the shore of oil spills. In addition, fertilisers were added to boost the growth of microbes, and oil on certain rocky surfaces vanished in just in 2–3 weeks.

Previously considered highly toxic and non-destructive, industrial pollutants like polychlorinated biphenyls (PCBs) are now successfully dechlorinated by this method.

2. Another method to achieve bioremediation at the polluted site is to remove microbial sample from the site, enrich the useful microorganisms, scale-up from the mixture by bioreactor cultivation and re-inoculate large quantities of these microbes into the contaminated site. The method has some constraints which are listed below:

1. Adaptation of indigenous microbes to specific environment of the contaminated site must be achieved.

2. When foreign microbes are introduced into the site, they must survive and compete with the well-adapted indigenous microorganisms in the hostile environment.

3. The mixture of microbes that is added to the site must remain in close contact with the pollutant to achieve degradation.

4. If the mixture contains indigenous microbes and genetically modified organisms (GMOs) to achieve speedy bioremediation, there is great risk of spread of gene-manipulated cultures to the environment.

Bioremediation is a technology that will require time for full development and application. Several companies, especially in the US, offer mixtures of microbes and immobilised enzymes designed to clean up chemical wastes, including oil, detergents, waste water from paper mills, etc. This technology will provide a means to improve many types of existing biological and chemical engineering processes that presently generate environmentally damaging by-products. In total, environmental biotechnology will play a central role in developing "Ecotech", a new technology concept.

Phytoremediaton With the advent of biotechnology it is possible to use plants that can clean up contamination. Plants that can grow on contaminated soil can be used to 'mop up' the heavy metal pollution from mining waste disposal and sewage sludge. By identifying the gene that enables the plant to withstand high levels of metal pollution it becomes possible to modify other plant species for use as environmental clean-up agents.

10.7 MICROBES AND MINING

The geological processes like formation and degradation of minerals, sedimentation, weathering and geochemical cycling are catalysed by microorganisms. There are several examples of microbes having detrimental effects on minerals including formation of acid mine waters, production of sulphuric acid, weathering of limestone, etc. Such microbial activities have caused the gold mines of Wales and California to be abandoned; in India, Taj Mahal is the most beautiful monument made of limestone undergoing defacement as a consequence of microbial action.

In spite of the fact that microbes produce several harmful effects, they are increasingly employed for extraction of commercially significant metals such as, copper, cobalt and uranium from rocks by solubilisation (bioleaching). Microbial mining is no more a fantasy.

Thiobacillus ferrooxidans is probably one of the oldest forms of life on earth. This bacterium is present in many types of rock throughout the world, and there can be many millions in just a handful of material. Usually it lives in total darkness, therefore it does not depend on sunlight or organic matter in the surrounding to obtain energy. Instead, it derives energy from inorganic compounds like iron sulphide and uses that energy to construct the materials it needs to live from the carbon dioxide and nitrogen in its environment. In the process it also manufactures sulphuric acid and iron sulphate. This ability of *Thiobacillus ferrooxidans* to oxidise both sulphur and iron is exploited in many mining operations.

The sulphuric acid and iron sulphate produced by *Thiobacillus ferrooxidans* attack the surrounding rocks and leach (dissolve) many metallic minerals. This process is called bioleaching. Due to activities of these microbes, insoluble copper sulphide is converted into soluble copper sulphate. As water percolates through the rocks, copper sulphate is carried along and eventually collects as a bright blue pool. This is the method employed in obtaining concentrated copper from a metal-rich lagoon, that was scattered throughout thousands of tonnes of low-grade ore. The copper is recovered by passing copper sulphate solution over pieces of iron. As a consequence a layer of copper is deposited on the iron that can be scraped off. Bioleaching of low-grade ore for uranium extraction is based on the same type of process.

Key Concepts

- Bioremediation involves biological methods to clean up contaminated soil, underground water, oceans, etc. and the process is also called biorestoration, bioreclamation or biotreatment.

- Catastrophic and the adverse effects of oil spillage on water ecosystem can be minimised by bioremediation.

- The rate of biodegradation by indigenous microbes can be accelerated by addition of nutrients to increase the growth *in situ.*

- The activity of microorganisms can also be accelerated by addition of a mixture of microbes at the polluted site. The bioremedial approach is successfully employed to clean up chemical wastes including oil, detergents, paper mill wastes, etc.

- The role of microbes in mineral mining is fabulous.

- In spite of the detrimental effects produced by microbes on minerals they are increasingly employed for extraction of commercially significant metals like copper, cobalt, lead, nickel and the precious metal uranium.

- Intellectual utilisation of the bacteria *Thiobacillus ferrooxidans* made the process of extraction of minerals from the mines economical.

- Worldwide extraction of copper and uranium from low-grade ores is based on bioleaching processes, thus microbes play an indispensable role in mineral mining.

In USA, almost 15% of total copper production is based on this biotechnology. Countries like India, Canada, USA, Chile and Peru are routinely extracting copper from low-grade ores by bioleaching, at the rate of 300000 tonnes per year. Uranium has a significant role to perform in energy production. Bioleaching of uranium ores have an important contribution to the economy of nuclear power stations by providing a way of recovery of uranium from low-grade nuclear wastes. This is in practice in countries like USA, Canada, India and the former Soviet Union.

Microbial mining has cut extraction costs by a quarter as compared to conventional mining methods. Semi-industrial processes have already shown the promise of microbial leaching in the recovery of cobalt, lead and nickel. Designing an organism by genetic engineering techniques for a specific function in the leaching process could be further beneficial in mining industries.

Review Questions

1. Define waste and pollutant. Explain the various sources of waste.

2. Explain different types of pollutants with their hazards.

3. Define xenobiotic compounds. Explain the role of different microorganisms in detoxification of xenobiotics.

4. Explain the various methods for management of solid, liquid and gaseous pollutants.

5. What is bioremediation? Explain different approaches to clean up an oil spillage site.

6. Describe the role of microbes in mineral mining with reference to copper extraction.

7. Write short notes on:

    a. Industrial pollutants

    b. House-building activities

    c. Biomedical wastes

    d. Hazards due to biological agents

    e. *Pseudomonas* in biodegradation

    f. Bioaugmentation

    g. Biomagnification

    h. Landfill

    i. Recalcitrant

    j. Biofilters

    k. Activated sludge treatment

    l. Bioleaching

11

RECOMBINANT DNA TECHNOLOGY
ENVIRONMENTAL, HEALTH AND SAFETY, SOCIO-ECONOMIC, AND ETHICAL CONSIDERATIONS

11.1 Introduction

11.2 Environmental issues

11.3 Health and safety issues

11.4 Socio-economic issues

11.5 Ethical issues

11.1 INTRODUCTION

Recombinant DNA technology or genetic engineering is the genetic modification of organisms by artificial methods. It has a major share in the success of biotechnology. The earlier chapters have elaborated the different techniques of genetic engineers to exploit the benefits in the form of

1. Increased food production.
2. Improved farm animals and their products.
3. Improved crops and ornamental plants.
4. Improved diagnostic tools.
5. Prenatal detection of genetic ailments.
6. Gene, enzyme or hormonal therapies for various genetic defects.
7. Novel vaccines to fight various diseases.
8. Bioindustrial revolution for production of enzymes and useful chemicals.
9. DNA profiling for criminal investigation.
10. Management of hazardous wastes to cut down environmental pollution.
11. Production of low-cost and ecofriendly biofuels.
12. Economical and safe microbial mining.
13. Xenotransplantation.
14. Understanding the structure and function of gene.
15. Human genome study.
16. Ambitious research to develop new therapies for cancer and AIDS.
17. Extension of research to develop new scientific disciplines—bioinformatics, proteomics, etc.

The list of the benefits is ever-expanding, but turning a brilliant idea into commercial usefulness that will benefit the society is never easy. The side effects of genetic engineering are associated with environmental, health and safety, socio-economic, and ethical issues.

11.2 ENVIRONMENTAL ISSUES

Since the 1970s a lot of public concern and debate have existed regarding genetic engineering experiments with microorganisms. Many molecular biotechnologists were in the belief that the process was unsafe and there should be strict regulations about the release of genetically manipulated organisms (GMOs) into the environment. The fundamental fact was that GMOs could escape from the manufacturing site into the environment and could cause unpredictable and catastrophic effects in the environment and ultimately disturb the balance in the ecosystem. There also existed a fear about the prolonged use of transgenic plants and animals which may lead to reduction in natural genetic diversity.

In response to these concerns, The Agriculture and Environment Biotechnology Commission (AEBC) and Advisory Committee on Release to Environment (ACRE) in UK have made a survey of public opinion regarding science and technology. In March 2000, the committees reported that there is a general level of support for science and technology among British people and it is also recommended by the committee that there should be constant dialogue between public concern, consumer organisations and science-based policy makers to evaluate the risk and benefits of various technologies. Such integrated approach can resolve the problem of risk and maximum benefits from different technologies can be exploited (figure 11.1).

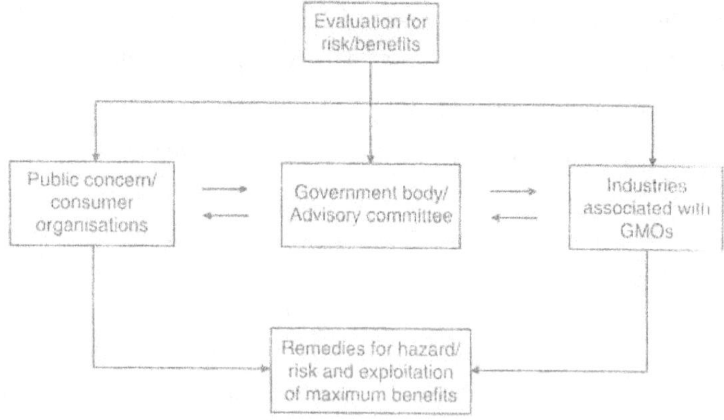

Figure 11.1   Scheme for assessment of risk and benefit of genetic engineering

Since 1987, there have been over 2000 field trials with transgenic plants in the USA under strict conditions that prevent the movement of plants and pollen from the test sites. Thus these plants behave like ordinary crop plants. The main concern with reference to transgenic plants is related to their cultivation in real field situation where there may be a possibility of their becoming persistent weeds, or they may transfer the gene to other plants creating ill effects in the ecosystem. The Department of Biotechnology, New Delhi, India is concerned with the regulation of field-testing of transgenic plants. Planned Release of Selected and Manipulated Organisms (PROSAMO) and ACRE, UK have carefully monitored and recorded the release of all GMOs into the environment. So far there are no reports of negative impacts on the ecosystem.

11.3 HEALTH AND SAFETY ISSUES

Recombinant DNA technology has shown some adverse effects on health proving its unsafe nature. On the basis of such effects observed in some cases, some of the projects have been abandoned and they require specific approval from regulatory authorities to continue their operations. A few such cases are listed below.

1. In 2003, two children suffering from severe combined immunodeficiency (SCID) in France received gene therapy to correct this disorder. But both developed leukemia. As a result, no more children in the UK will be given the gene therapy treatment for SCID without specific approval of the UK regulatory authority, on a case-by-case basis.

2. Similarly, the release of synthetic l-tryptophan, which was produced by the Japanese sector using genetically engineered *Bacillus amyloliquefaciens,* resulted in eosinophilia-like symptoms, and this was also banned.

3. In USA, a number of children, treated with growth hormone produced by rDNA technology are suffering from severe health hazards indicating that the product is unsafe.

4. The BT endotoxin gene introduced in many of the crop plants showed stunted growth.

5. The food derived from genetically modified microbes, plants or animals induced allergic responses and created health

problems. In addition, transferred antibiotic-resistance gene into tomatoes, melons, and berries are likely to be transferred to human beings who consume them.

In USA, UK and Japan, the government health and safety authorities supervise and control rDNA work. The Genetic Manipulation Advisory group (GMAG), the Advisory Committee on Novel Food and Processes (ACNFP) and the Gene Therapy Advisory Committee (GTAC) have established procedural guidelines for confinement of rDNA activities within the laboratory. After evaluating the benefits and risks of the product to be released, they reserve the rights for approval of license to the manufacturing sector. The committees are influenced only by the scientific facts and ultimate safety of the product. At the same time public confidence must also be achieved for the success of any new biotechnological product.

However, in the light of global events, biomedical research that involves the use of potentially harmful pathogens and toxins has come under increased scrutiny, and there are heightened concerns that the misuse of this research could increase the potential threat of bioterrorist attacks.

The Wellcome Trust, UK, recognises that there are particular concerns regarding research that could directly result in, or enable the future development of pathogens and toxins which could potentially serve as bioweapons. A committee convened by the US National Academy of Sciences recently identified seven classes of experiments that would require careful review by informed experts. The experiments that this committee specified are those that would

- demonstrate how to render a vaccine ineffective;
- confer resistance to therapeutically useful antibiotics or antiviral agents;
- enhance the virulence of a pathogen, or render a non-pathogen virulent;
- increase transmissibility of a pathogen;
- alter the host range of a pathogen;
- enable the evasion of diagnostic and detection modalities; or
- enable the weaponisation of a biological agent or toxin.

The Trust considers that in order to address these legitimate concerns, it is important that appropriate processes exist at institutional, national and international levels for the review of research that could result in such outcomes.

The Trust would emphasise, however, that further research involving harmful biological pathogens and toxins will be crucial in the fight to combat the diseases that these agents cause and to improve our ability to respond to bioterrorist attacks. In most cases the risks associated with such research will be minor in comparison with the potential benefits. The Trust considers that the creation and dissemination of scientific knowledge is a definite and tangible public good, which would need to be set against risks which may sometimes be hypothetical and hard to quantify.

The Trust believes, therefore, that regulatory processes must not unduly restrict this essential research. Any additional regulatory requirements that may be introduced should apply only to those research projects where there is tangible cause for concern. The Trust expects that this will represent a very small proportion of the many research projects undertaken in academic research laboratories both in the UK and at centres of excellence in developing countries that involve the use of pathogens and toxins.

11.4 SOCIO-ECONOMIC ISSUES

For many aspects of new biotechnology, there is a social price paid by the people living in developing countries.

1. The novel sweeteners in the form of saccharine, aspartame (nutra-sweet) and thaumatin are 1000,000 times sweeter than sugar. These substituted sweeteners are already reducing the traditional sugar market for sugar cane and sugar beet, disrupting these economics.

2. Increased milk production from fewer cows by the injection of genetically engineered hormones will result in many small farmers in the USA and EU being put out of business.

3. In some countries, more of investment is made for research in the combating of rare genetic disorders than for research in a more common malarial vaccine. rDNA technology is employed

on potato to produce plastic (bioplastic) from starch but there is a scarcity of potato as food.

4. The multinational agrochemical companies which are engaged in the production of genetically engineered crops and seeds will have to recover the cost of their high research and investment only from the farmers. Farmers having sound economic background can bear this cost burden, but it will not be possible for the farmers living in the third world.

5. Some of the traditional farming practices and disease treatment methods are on the verge of extinction due to the advent of genetic engineering. This may be a cause of concern particularly in the context of unemployment. However, the Office of Technology Assessment in USA, the Institute of Manpower Studies in the UK and the Organization for Economic Cooperation and Development (OECD) reported that the growth of biotechnology will not substantially affect total employment in the short term. Looking further into the future, however, it is possible to foresee more jobs being created for existing products and manufacturing methods, as biotechnology yields entirely novel products rather than replacement, directly or indirectly.

11.5 ETHICAL ISSUES

Recombinant DNA technology is now being extensively used to improve specific characteristics of plants and animals for various purposes, including DNA fingerprinting, human genome study, patenting, cloning etc. which have created several moral and ethical problems.

1. Indiscriminate use of women in reproduction and genetic screening (including amniocentesis) is an exploitation of women.

2. Introduction of human gene into animals and vice-versa and inserting plant genes into microorganisms and other species raised several moral and ethical questions.

3. Patenting genetically engineered animals is equating them to automobiles or mere factories. Animals are also living beings, which feel pleasure and pain just as we do.

4. The diagnostic procedures including DNA fingerprinting, amniocentesis, etc. undermine individual privacy.

5. Genetic modification or cloning of human beings should not be attempted.

6. The Committee of the Ethics of Genetic Modification and Food Use in the UK, reported some of the ethical concerns related to the use of food in certain transgenic organisms, which are listed below.

   a. Transfer of human genes to animals used as food.

   b. Transfer of genes from animals whose flesh is forbidden for use as food to animals that are normally eaten would offend certain religious groups like Jews and Muslims.

   c. Transfer of animal genes into food plants may be of particular concern to some vegetarians.

   d. Use of organisms containing human genes as animal feed.

The food products derived from transgenic organisms containing copy genes that are ethically unacceptable to some groups of the population subject to dietary restriction by their religion should be so labelled, to help the selection. The report strongly recommended that whenever the alternatives could be found, the use of ethically sensitive genes in food organisms should be discouraged. The committee also reported that the chances of recovering the original human gene from the transgenic embryo are much less than the chances of recovering a specific drop of water released into an ocean, since the transgenic organism does not contain the actual human gene but rather an artificially created copy of the gene.

In relation to prenatal diagnosis, perhaps the most worrying aspect is the severely disturbed sex ratio, particularly in developing nations. This is because there are several cases of aborting the female foetus. Though there are legal restrictions on prenatal detection of sex of the baby, it is an impossible task to monitor and control it.

In the context of gene therapy, from an ethical point of view, somatic gene therapy should be used only to alleviate serious medical disorders and not for non-therapeutic applications. It must remain under close supervision to satisfy medical safety, legal implication and public concern. On the other hand, germ line gene therapy is extremely difficult and it is ethically and socially unacceptable.

Interfering with germ cells raises problems of eugenics and there must be public debate before its application.

11.5.1 Consequences of Cloning

Another anxiety and fear-producing aspect of genetic engineering is the cloning of animals and man. Although there are possible benefits of successful cloning, potential for misuse is real. In 1997, Dr. Wilmut created Dolly the cloned sheep, which died in 2003, leaving behind several problems including the age factor. In the cloning experiments of Dolly, the original cell used for making its photocopy, was from a six-year old sheep that might have accumulated many mutations and the sheep lived shorter than the parent because it started from the parent cell which was already six year old. Trials are on to produce a clone of the most powerful human but this will ultimately have some defects as far as its lifespan is concerned. In March 2004, a team of genetic engineers headed by Dr. Wilmut announced successful cloning of human embryo but they pointed out that this experiment is limited to the few-celled stage only. There are several annoying issues in relation to cloning. After much of the debate between scientists, policy makers and governmental organisations, in August 2004, the UK Government has authentically permitted cloning of human embryo that will be restricted to essential research.

1. Man will place himself at God's position, since man becomes a creator of human beings.

2. Cloning of humans is an inherently evil act.

3. The relationship between man and woman will be meaningless.

4. In future there will be no 'family' concept.

5. Children will be considered as products only.

6. Women will be considered as incubators available on rent.

7. Producing photocopies of single individual will become meaningless and will be injustice to new identical ones.

8. Biodiversity on this planet will be lost.

Are all these issues real or imagined? Time will provide the answer. Dr. Wilmut comments on the outcome of genomic research, thus— "You cannot blame the scientists for making these kind of discoveries. It is upto society to decide how it should be used". The public do not accept or reject gene technology as a whole. Parts of it will be welcomed and utilised while others will have less or no support at all. In UK a workshop on ethical issues related to biotechnology was conducted in April 2002 in the presence of the Ethics and Governance Council (EGC) which ended with the following message to biotechnology companies: "Provide the information and listen to the public." All in all, the technologies should not be blamed but their misuse should be controlled. After much of the debate between scientists, policy makers and governmental organisations, in August 2004, the U.K. Government has authentically permitted cloning of human embryo that will be restricted to essential research.

Key Concepts

- Genetic engineering has made the transfer of gene from an organism to any other organism technically feasibile.

- Recombinant DNA technology has a major impact on health and pharmaceutical industry and in agriculture, food and environment.

- The development of new drugs, human and animal vaccines, diagnostics, transgenic crops and animals and a wide range of new processes to clean up and manage hazardous waste must undoubtedly bring substantial improvement in the living standards of people.

- In spite of the tremendous benefits of rDNA technology, there are several environmental, health and safety, socio-economic and ethical issues related to genetic engineering.

- All these issues suggest there is an increasing awareness among the people, governmental and non-governmental bodies, industry, consumer organisations, etc.

- In spite of the obvious difficulties, genetic engineering and biotechnology will have an increasingly significant role to perform in our society.

Review Questions

1. Explain the role of genetic engineering in the improvement of the life of human beings.

2. What are the different issues raised in relation to recombinant DNA technology.

3. Write short notes on:

    a. Advisory committees

    b. Assessment of risk and benefits

    c. Ethical issues related to cloning

APPENDIX-I

CHRONOLOGICAL EVENTS IN THE FIELD OF GENETICS

1865   G.Mendel presents principles of heredity.

1869   Chemical material discovered in the cell.

1871   Friedrich Miescher calls the nuclear material as "nuclein".

1879   Walter Fleming discovers chromatin, now called chromosome.

1889   Richard Altman coins the term "nucleic acid".

1899   First International Congress of Genetics is held in London.

1905   William Bateson coins the word "Genetics".

1909   The word "Gene" first used.

1919   The word "biotechnology" first used.

1920   Chromosomes proposed to be the carriers of inherited characteristics.

1928   Frederick Griffith identifies transforming principle in bacteria.

1940   Hans Spemann divides fertilised frog egg into two parts.

1941   The term "genetic engineering" first used.

1944   Oswald Avery identifies DNA as genetic material.

1951   E.Chargaff shows how DNA's four chemical bases pair up.

1952   Harshey and Chase prove that DNA carries genetic information.

1952   R.Briggs and T.J.King conduct nuclear transplant experiment.

1952   Franklin and Wilkins conduct X-ray crystallography of DNA.

1953   Watson and Crick deduce DNA structure.

1961   Genetic code cracked by H.G.Khorana and M.Nirenberg.

1970   John Gurdon's first successful cloning of tadpole.

1973   Cohen and Boyer produced first recombinant gene from bacteria.

1975   F.Sanger and A.R. Coulson's DNA sequencing method.

1981   First cancer-causing gene identified.

1982   Rat growth hormone inserted into mouse eggs to produce the first supermice.

1982   Humulin—the first genetically modified insulin drug introduced.

1984   Alec Jeffreys discovers DNA profiling.

1984   McGrath and Solter conduct nuclear transplant experiment in mouse.

1986   The field test of genetically modified plant.

1986   The first genetically modified human vaccine goes into clinical use.

1986   Steen Willandsen—splitting of fertilised sheep eggs to develop identical lambs.

1988   Neal First and colleagues carry out fusion between cells from cattle embryos and enucleated cattle eggs.

1990   Human genome project launched.

1990   Transgenic sheep, Tracy, born in Scotland.

1993   The gene for Huntington's disease identified.

1993   Gene therapy for severe combined immune deficiency (SCID) and cystic fibrosis begins in UK.

1993   Hall Stillman's experiment on artificial splitting of embryos of human beings to produce identical twins.

1996   The gene for Parkinson's disease identified.

1997   Don Wolf *et al* carry out nuclear transplant from embryos to eggs in monkeys.

1998   First complete genome of a multicellular organism, the nematode worm *C.elegans,* sequenced.

2000   Draft sequence (90%) of human genome announced.

2000   Golden rice modified to make vitamin A to protect against blindness.

2000   Genome of fruit fly *Drosophila* sequenced.

2000   First complete genome of a plant, *Arabidopsis thaliana,* sequenced.

2001   First transgenic piglets offer possibility of xenotransplantation for organ transplant patients.

2002   Rice becomes the first food crop to have its genome decoded.

2002   First public draft of genome of mouse, important model animal, announced jointly by UK and US institutions.

2003   Completion of human genome project.

2004   Dr. Wilmut makes an announcement regarding the cloning of
(Mar) human embryos.

2004   U.K. Government permits scientists to carry out experiments in
(Aug) human cloning.

APPENDIX-II

NOBEL PRIZE AWARDED FOR PHYSIOLOGY, MEDICINE AND CHEMISTRY

Year	Scientist/s	Research
2003	Paul.C. Lauterbur, and Sir P. Mansfield	Magnetic resonance imaging (MRI).
2002	Sydney Brenner, H. Robert and John E. Sulston	Genetic regulation of organ development and programmed cell death.
2001	Leland H. Hartwell, R. Timothy and P. M. Nurse	Key regulators of the cell cycle.
2000	Arvid Carlsson, Paul Greengard and Eric Kandel	Signal transduction in the nervous system.
1999	Gunter Blobel	Proteins have intrinsic signals that govern their transport and localisation in the cell.
1998	Robert F. Furchgott, Louis J. Ignarro and Ferid Murad	Nitric oxide as a signaling molecule in the cardiovascular system.

Year	Scientist/s	Research
1997	Stanley Prusiner	Prions—a new biological principle of infection
1996	P. C. Doherty and R. F. Zinkernagel	Specificity of the cell-mediated immune defence.
1995	E. B. Lewis, C. N. Volhard and E. F. Wieschaus	Genetic control of early embryonic development.
1994	Alfred G. Gilman and M. Rodbell	G-proteins and the role of these proteins in signal transduction in cells.
1993	R. J. Roberts and P. A. Sharp	Split genes and RNA processing.
1993	K. Mullis	Polymerase Chain Reaction.
1993	M. Smith	Site-directed mutagenesis.
1992	E. H. Fischer and E. G. Krebs	Reversible protein phosphorylation as a biological regulatory mechanism.
1991	E. Neher and B. Sakmann	Function of single ion channels in cells.
1990	J. E. Murray and E. D. Thomas	Organ and cell transplantation in the treatment of human disease.
1989	J. M. Bishop and H. E. Varmus	Cellular origin of retroviral oncogenes.
1989	T. R. Cech, S. Altman	Ribozyme.
1988	Sir J. W. Black, G. B. Elion and G. H. Hitchings	Important principles for drug treatment.
1988	J. Deisenhofer, R. Huber and H. Michel	Bacterial photosynthetic reaction centre.
1987	Susumu Tonegawa	Genetic principle for generation of antibody diversity.

Year	Scientist/s	Research
1986	Stanley C. and Rita Levi-Montalcini	Growth factors.
1985	M. S. Brown and J. L. Goldstein	Regulation of cholesterol metabolism.
1984	N. K. Jerne, G. J. F. Kohaler and Cesar Milstein	Specificity in the development and control of the immune system and the discovery of the principle for production of monoclonal antibodies.
1983	Barbara Mc Clintock	Mobile genetic elements.
1982	John R. Vane	Prostaglandins and related biological active substances.
1982	A. Klug	Structure of virus and components of life.
1981	David H. Huble and T. N. Wiesel	Information processing in the visual system.
1981	Roger W. Sperry	Functional specialisation of cerebral hemisphere.
1980	Baruj B., J. Dausset and G. D. Snell	Genetically determined structures on the cell surface that regulate immunological reactions.
1980	P. Berg, W. Gilbert and F. Sanger	DNA sequencing technology.
1979	A. M. Cormack and Sir G. N. Hounsfield	Computer assisted tomography, CAT-scan
1978	Warner A., Daniel N. and Hamilton O. Smith	Restriction enzymes and their application to problems of molecular genetics.
1977	R. Guillemin and Andrew V. Schally	Peptide hormone production of the brain.
1977	Rosalyn Yalow	Radioimmunoassays of peptide hormones.

Year	Scientist/s	Research
1976	B. S. Blumberg and D. C. Gajdusk	New mechanisms for the origin and dissemination of infectious diseases.
1975	D. Baltmore, R. Dulbecco and H. M. Temin	Interaction between tumour viruses and the genetic material of the cell.
1974	A. Claude, C. De Duve and G. E. Palade	Structural and functional organisation of the cell.
1973	K. V. Frisch and Nikolaas T.	Organisation and elicitation of individual and social behaviour patterns.
1972	G. M. Edelman and R. R. Porter	Chemical structure of antibodies.
1971	E. W. Jr. Sutherland	Mechanism of the action of hormones.
1970	Sir B. Katz U. V. Euler and Julius Axelford	Humoral transmitters in the nerve terminals and the mechanism of release and inactivation.
1969	Max Delbruck, A. D. Hershey and S. E. Luria	Replication mechanism and the genetic structure of viruses.
1968	H. G. Khorana, R. W. Holley and M. W. Nirenberg	Interpretation of the genetic code and its function in protein synthesis.
1967	R. Granit, H. K. Hartline and George W.	Primary physiological and chemical visual processes in the eye.
1966	Preyton Rous	Tumour inducing viruses.
1966	C. B. Huggins	Hormonal treatment of prostatic cancer.
1965	F. Jacob, Andre L. and J. Monod	Genetic control of enzyme and virus synthesis.

Year	Scientist/s	Research
1964	K. Bloch and F. Lynen	Mechanism and regulation of cholesterol and fatty acid metabolism.
1963	Sir John C. E., Sir A. L. Hodegkin and Sir. A. F. Huxley	Ionic mechanisms involved in excitation and inhibition in the peripheral and central portions of the nerve cell membrane.
1962	Francis H. C. Crick, J. D. Watson and M.H.F.Wilkins	Molecular structure of nucleic acids and its significance in information transfer in living material.
1961	G. V. Bekesy	Physical mechanism of stimulation within the cochlea.
1960	Sir F. M. Burent and Sir Peter B. Medawar	Acquired immunological tolerance.
1959	S. Ochoa and A. Kornberg	Mechanisms in the biological synthesis of ribonucleic acid and deoxyribonucleic acid.
1958	F. Sanger	Primary structure of protein.
1958	G. W. Beadle and E. L. Tatum	Genes act by regulating definite chemical events.
1958	Joshua Lederberg	Genetic recombination and organisation of the genetic material of bacteria.
1957	Daniel Bovet	Synthetic compounds that inhibit the action of certain body substances, especially their action on the vascular system and the skeletal muscles.

GLOSSARY

abiotic Referring to the absence of living organisms.

acellular Lacking cellular organisation; not having a delimiting cytoplasmic membrane; organisational description of viruses, viroids and prions.

acidic A solution having pH less than 7.0. It releases hydrogen ions when dissolved in water.

AIDS An infectious disease syndrome caused by HIV retrovirus, characterised by loss of normal immune response followed by various opportunistic infections.

acquired immunity The ability of an individual to produce specific antibodies in response to antigens to which the body has been previously exposed based on the development of memory response.

activated sludge process It is an aerobic secondary sewage treatment process using aerobic microorganisms present in sewage sludge to break down organic matter in sludge.

adenine One of the four nitrogen bases that make up the letters A, T, G and C in DNA code.

Adenosine Triphosphate (ATP) An energy-rich compound present in biological systems, composed of adenosine and three phosphate groups. The free energy released from the hydrolysis of ATP is utilised to carry out energy-dependant processes in the biological systems.

aerobic Living or acting only in the presence of oxygen.

agitate To move a solution with rapid motion in order to maximise the mixing of tissue or cell and nutrient medium inside the fermenter.

Agrobacterium tumefaciens The bacterium causing crown gall disease of plants. The Ti plasmid of *A. tumefaciens* is known to cause the disease and a small portion of it is used as a vector in the genetic modification of plants.

alkaline   A basic solution with a pH above 7.0.

allele   One of the many possible forms of the gene; different alleles produce variation in the inherited characteristics.

allosteric enzyme   Enzyme with a binding and catalytic site for the substrate and a different site where a modulator (allosteric effector) acts.

amino acid   The building blocks of protein.

amniocentesis   Puncture of the uterine wall with a needle for the purpose of obtaining amniotic fluid, which can be analysed to determine whether the foetus has a genetic abnormality.

amplification   Production of additional copies of a chromosomal sequence.

anaerobes   Microorganisms that can grow and multiply in the absence of oxygen.

antibiotic   One of many natural organic substances (or their synthetic analogues) secreted by undergrowth or microbes that are toxic to other species, retarding or preventing their growth and this presumably functions as a defence mechanism.

antibodies   Proteins produced by the immune system in order to neutralise potentially harmful foreign molecules (antigens).

antigen   A substance that can be recognised by the body as alien and that may induce an immune response by antibody.

antisense therapy   The *in vitro* treatment of the genetic disease by blocking translation of a protein with a DNA or RNA sequence that is complementary to a specific mRNA.

artificial seed   Encapsulated or coated somatic embryos (embryoids) that are planted and treated like seed.

assay   1) To test or evaluate. 2) The substance to be analysed or the process of examining or testing it (chemically or by other means).

autoimmunity   A situation in which the body produces an immune response against its own organs or tissues.

autosome   Any chromosome that is not a sex chromosome; human somatic cells have 22 pairs of autosomes and one pair of sex chromosomes.

autoradiography   Technique by which emissions from radioactively labelled microscopic specimens can be visualised by exposure of the labelled tissue to photographic emulsion.

bacteriophage   A virus that infects the bacteria. It replicates in bacteria. In genetic engineering it is used to introduce genes into the bacterial cell.

bacterium   A structurally simple single cell with no nucleus. Its role in biotechnology is indispensable.

base pair (bp)   A pair of molecules (either A and T or G and C) present on complementary strands of double-stranded DNA. Pairing is also possible in RNA under some circumstances.

batch culture   A cell suspension grown in liquid medium of a set volume.

B lymphocyte   White blood cells that are able to produce specific antibodies.

bioinformatics   The application of computer and statistical techniques for the management of biological information which plays an important role in drug designing.

bioleaching The use of microorganisms to transform elements (minerals) so that they can be extracted from material when water is filtered through it.

biomass All organic matter that are derived from the photosynthetic conversion of solar energy.

bioreactor (fermenter) A device or container designed for fermentation.

bioremediation The use of microorganisms to clean up hazardous wastes in the environment.

biosensor An electronic device that uses biological molecules to detect specific compounds.

biotechnology The industrial use of biological process.

blood serum The fluid expressed from clotted blood or clotted blood plasma.

callus Unorganised plant cell mass capable of repeated cell division and growth *in vitro*.

cation An ionised atom with a net positive charge.

cell culture The technique used to grow cells outside the organism.

cell fusion/hybridisation Technique whereby two different types of cells (from one parent or from different species) are joined to produce one cell with one continuous plasma membrane.

cell line Cells that have acquired the ability to multiply indefinitely *in vitro*.

cellulose Complex polysaccharide forming the cell wall.

chimera A tissue with cells of more than one genetic make-up.

cistron/gene The functional unit of genetic inheritance; a segment of genetic nucleic acid that codes for a specific protein.

clone A collection of genetically identical cells or organisms derived from a common ancestor.

cDNA (complementary DNA) DNA strand formed from RNA using the enzyme reverse transcriptase.

continuous fermentation A fermentation process that can run for long periods and in which the raw material is supplied and the products and microorganisms are removed continuously.

cloning vector Any plasmid or phage into which foreign DNA to be cloned may be inserted.

coenzyme An organic, non-protein component of an enzyme.

cofactor The nonprotein component of an enzyme, which can be either inorganic or organic in nature.

codon A group of three nucleotides of mRNA coding for an amino acid.

cohesive ends DNA with single-stranded ends that are complementary to each other, helping the molecules to join. Bacteriophage lambda has cohesive ends (the 'cos' site) which allows formation of concatemers.

colony hybridisation A technique for using *in situ* hybridisation to identify bacteria carrying chimeric vectors whose inserted DNA is homologous with some particular sequence.

conjugate protein Protein linked covalently or noncovalently to substances other than amino acids, such as metals, nucleic acids, lipids, and carbohydrates.

cosmid Plasmid into which phase lambda cos sites have been inserted. As a result, the plasmid DNA can be packaged *in vitro* in phase coat. This vector is designed to clone large fragments of eukaryotic DNA.

covalent bond The type of chemical bond in which electron pairs are shared between two atoms.

crown gall A disease of plant in which tumours are formed due to infection caused by *Agrobacterium tumefaciens*. A tumour-inducing portion of bacterial genome (the Ti plasmid) may be used experimentally as a genetic tool to incorporate genetic information into plant cells.

cybrid A fusion product arising out of fusion between protoplast containing a nucleus and protoplast without nucleus.

cytosine One of the nitrogen bases found in nucleic acid.

dalton Unit of molecular weight; one dalton equivalent to one unit of atomic mass (e.g. the mass of a ^{1}H atom)

dehydrogenase An enzyme that removes hydrogen.

denaturation of DNA Conversion of double-stranded DNA to single-stranded state usually by heating.

DNA cloning A broad base technique for producing large quantities of a specific DNA segment.

DNA libraries Collection of DNA fragments representing the entire genome of an organism.

DNA polymerase Enzyme responsible for synthesis of a new DNA strand.

DNA probes Isolated single DNA strand used to detect the presence of complementary sequences.

DNA profiling The process of identifying the genetic make-up of an individual.

DNA tumour viruses Viruses capable of infecting vertebrate cells, transforming them into cancer cells. These viruses have DNA in mature virus particle.

de novo Arising from very simple precursors.

direct embryogenesis Embryoid formation directly on the surface of zygotic or somatic embryos in culture without callus formation.

disulphide bond A covalent bond formed between –SH groups of two cysteine residues (-S-S-).

downstream processing Separation and purification of product/s from a fermentation process.

effluent The liquid discharge from sewage treatment and industrial plants.

embryoid Embryo-like structure derived from the cells of the vegetative plant body grown *in vitro*.

embryo transfer Implantation of embryos obtained from donor animals or generated by *in vitro* fertilisation into the uteri of recipient animals.

endonuclease Enzyme that catalyses the cleavage of DNA, and normally cuts DNA at specific sites.

endotoxin Toxic substances found as part of some bacterial cells.

enzyme   A class of proteins that control biological reactions.

Erlenmeyer flask   A wide-necked variety of conical, flat-bottomed flasks commonly used for medium preparation.

eugenics   Application of the laws of genetics to the methods employed in the improvement of human race. It deals with the science of well-born, improving the inborn qualities of the race and obtaining a better heritage by judicious breeding.

eukaryotes   Cellular organisms having a membrane-bound nucleus in which the genome of cell is stored. They include algae, fungi, protozoa, plants and animals.

explant   A part of the entire plant used for *in vitro* culture.

fermentation   The process by which microorganisms turn raw material into useful products.

fibroblast   Flattened connective tissue cell.

flocculation   A method by which an aggregation of microbes floating in or on liquids is separated.

functional genomics   A wide-ranging set of biological techniques used to identify or confirm the biochemical role of the gene.

fungi   A group of diverse, widespread, unicellular and multicellular eukaryotes lacking chlorophyll and usually bearing spores and often filaments.

fusogen   A fusion-inducing agent used for protoplast agglutination in somatic hybridisation studies, e.g. polyethylene glycol.

gene   The fundamental physical and functional unit of heredity. A gene is a sequence of DNA that codes for protein, enzyme or RNA. A virus such as HIV has about a dozen genes, bacteria can have about 5000 genes, yeasts can have about 7500 genes and humans have around 30000 genes.

gene amplification   Process by which a gene is replicated selectively.

gene bank   Collection of cloned gene fragments from a single genome.

gene therapy   The process by which a patient is cured by altering his or her genotype.

genetic code   The DNA sequence of a gene, which determines the sequence of amino acids in a protein or enzyme.

genetic engineering   The technology used to isolate genes from an organism, manipulate them in the laboratory and insert them into another cell system for industrial, medical and research purposes.

genetic mapping   Determination of relative positions of genes in DNA or RNA.

gene expression   The process in which the gene is activated and the DNA code of that gene is translated into specific protein.

genomics   The use of modern high throughput methods to study and analyse the whole genome in order to identify gene function.

genomic library   A set of cloned fragments together in vectors representing the genome.

genotype/genome   The genetic make-up of an organism or individual.

Genetically Modified Organisms (GMO)   An organism whose genetic make-up has been changed by any method including natural processes and genetic engineering.

guanine   One of the four nitrogen bases found in nucleic acids.

HeLa cell   Cell line originally derived from uterine carcinoma cells of a patient named Henrietta Lacks.

heterokaryon   A single cell containing nuclei of two types produced by fusion of two cells.

heterozygous   Having two different alleles for a given trait.

Hindenburg   A German airship completed in 1936 destroyed by fire with heavy loss of life while landing at Lakehurst, N.J., in 1937.

homozygous   Having two identical alleles for a given (gene in the diploid cell) trait.

Human Genome Project   An international research effort begun since 1980 to map and sequence all DNA found in human beings.

hybrid   The offspring formed by mating two plants/animals that differ genetically.

hybridoma   Hybrid cell line produced by fusion of a normal lymphocyte with a myeloma cell that produces quantities as specific antibodies.

immobilised enzyme   An enzyme whose free movement is restricted usually by entrapment, adsorption, encapsulation or covalent bonding with the help of polymer matrix.

incompatibility   Selectively restricted mating competence that limits fertilisation or physiological interaction resulting in rejection of a graft.

indirect embryogenesis   Embryoid formation on callus tissue derived from zygotic or somatic embryos, seedling plant or other tissues in culture.

inoculum   A material introduced onto or into a host or medium.

in situ hybridisation   Technique in which DNA is kept in place while it is allowed to react with particular preparation of labelled DNA or RNA.

intron   Intervening sequence of DNA located within a gene, that is not included in the mature mRNA.

in vitro   Literally means "in glass", e.g. a test tube. Experimentation on organisms or portions thereof in glassware or culture container under artificial conditions.

in vivo   Experimentation on organisms under natural conditions within intact living organisms.

isozyme   Multiple molecular forms of an enzyme exhibiting similar or identical catalytic properties. Their analysis involves electrophoresis and may confirm the production of somatic hybrid.

knockout mice   Mice, born as a result of a series of experimental procedures, which are lacking in a functional gene that they would normally contain.

lactose   A disaccharide of glucose and galactose found in milk.

ligase   Enzyme that joins cut ends of DNA strands.

lipofection   Delivery into eukaryotic cells of DNA, RNA or other compounds that have encapsulated in an artificial phospholipid vesicle (liposome).

locus The normal positions of gene on a chromosome.

lyophilise or freeze dry To freeze rapidly then dehydrate under high vacuum. The process is called lyophilisation.

metabolite Product of biochemical activity.

microbes Microscopic organisms that may be pathogenic or nonpathogenic.

microinjection The injection of DNA into a cell using a very fine needle.

monoclonal antibodies Antibodies that are derived from only one cell and that recognise only one type of antigen. They can be used for protection against disease causing organisms.

mutation The stable change in gene or chromosome that can cause a disorder or the inherited susceptibility to a disorder.

oncogene Gene thought to be capable of producing cancer.

origin The nucleotide sequence at which DNA synthesis is initiated. It is also called as origin of replication (ori).

organ culture The growth in aseptic culture of plant organs such as roots or shoots.

palindromic sequence Complementary DNA sequences that are the same when each strand is read in the same direction (e.g. 5' to 3').

parasexual hybrids Hybrids derived through fusion of protoplast from the same or different species.

pH The standard measure of relative acidity. It is a negative logarithm of the hydrogen ion concentration in any solution.

photon A packet of light energy.

phagocytosis Ingestion of particulate material by cells either as a means of feeding or defence.

plasmid An autonomous, self-replicating, extra-chromosomal DNA found in bacteria playing significant role in genetic engineering.

plasma cell Cells that are able to synthesise a specific antibody and secondary B-cells.

prokaryote Organisms that lack a true nucleus.

protoplast The organised living unit of a cell without a cell wall.

Poly A Long polyadenylic acid segment added post-transcriptionally to many eukaryotic mRNA at the 3' end.

polymer Large molecules made up of repeating structural subunits.

prion An infectious agent associated with certain mammalian degenerative diseases that are composed solely of protein; derived from *pro*teinacious *in*fectious particle.

protein Structurally and functionally diverse groups of polymers built of amino acid monomers.

protoplast Microbial or plant cell whose wall has been removed so that the cell assumes a spherical shape.

purine Type of nitrogenous bases found in nucleic acids having two fuse rings consisting of nitrogen and carbon atoms (e.g. adenine or guanine).

pyrimidine Type of nitrogen bases found in nucleic acids having single heterocyclic ring (e.g. cytosine, thymine or uracil).

quorn A fungus-derived protein that is gaining an increasing market place as a vegetable protein with meat-like texture and appearance.

recombinant DNA Genetic material with novel gene sequences produced by crossovers, chromosome reassortment, by other natural means or through genetic engineering.

replica plating A technique by which various types of mutants can be isolated from a population of bacteria grown under nonselective conditions, based on plating cells from each colony onto multiple plates and noting the position of inoculation.

restriction enzymes DNA endonucleases that can recognise specific short sequences of usually unmethylated DNA and cut the duplex DNA.

RNA tumour viruses Virus capable of infecting vertebrate cells, transforming them into cancer cells. RNA viruses have RNA in the mature virus particle.

scale-up Implementation of laboratory experiments to full size industrial process.

sendai virus Parainfluenza virus that adheres to cell surface and facilitates cell fusion.

site-directed mutagenesis A technique to modify a gene in a predetermined way so as to produce a protein with a specifically altered amino acid sequence.

somatic Refers to the vegetative or nonsexual stages of life cycle.

somatic embryogenesis Formation of embryos from asexual cells.

somatic hybridisation A technique of fusing protoplasts from two contrasting genotypes for production of hybrids or cybrids.

southern blotting Procedure for transferring denatured DNA from an agarose gel to a nitrocellulose paper where it can be hydrolysed with a complementary nucleic acid.

splicing Gene splicing or manipulation, the objective of which is to attach one DNA molecule to another. RNA splicing is the removal of introns from precursors (hn RNA).

stem cell An undifferentiated cell from which specialised cells develop.

sticky ends Single-stranded ends left on a restriction fragment by action of restriction enzymes.

thymine One of the four nitrogen bases that make up the letters ATGC in DNA.

transcription Formation of an RNA from DNA template with the help of RNA polymerase.

transgenic plants/animals The plants or animals that have been genetically engineered so that their chromosomes contain foreign genes.

transfection The process that introduces DNA into cultured cells.

transformed Normal cells that have been converted into cancer cells by treatment with a carcinogenic material, radiation, or an infective tumour virus.

translation Synthesis of protein in the cytoplasm using the information coded on mRNA.

tissue culture A process where individual cells, or clumps of plant or animal tissue are grown *in vitro*.

tumour-suppressor gene   Gene that encodes protein that retains cell growth and prevents cells from becoming malignant.

uracil   One of the four nitrogen bases that make up the letters AUGC found in RNA.

vaccine   A preparation that contains an antigen made up of disease causing organisms in a dead or weakened state. It is used to boost immunity against the given disease, and can be created using recombinant DNA technology.

vector   An entity, such as plasmids, viruses or bacteriophages, used to carry new DNA into an organism. In relation to parasitology, vectors like mosquitoes, flies, etc. are carriers of pathogenic organisms.

virus   A submicroscopic organism consisting of a DNA or RNA core surrounded by a protein coat capable of replicating the cells of the host organism.

xenotransplantation   The process of transplanting organs, tissues or cells from animals to humans.

x-ray crystallography/diffraction   Scattering of X-ray by the atoms of a crystal in a manner such that the resulting diffraction pattern provides information about the structure and/or identity of the substance.

SUGGESTED READINGS

A C R E, 2003, A report of Advisory Committee on Releases of genetically modified organisms to the Environment, U K.

Anon, 1993, Report to the committee on Ethics of genetic modification and food use, H M S O Publishing Centre, London.

BIOTAL (Biotechnology by Open Learning), A series of books, Butterworth Heinemann Ltd., Oxford.

Brinster, R. L., 1993, "Stem cells and transgenic mice in the study of development," *Int. J. Dev. Bio.,* 37:89–99.

British Council, 2003, DNA and After, 50 years of UK Excellence published by Foreign and Commonwealth Office, London.

Bu'lock, J.D. and Kristianusen, B., 1987, *Basic Biotechnology,* Academic Press, London.

Burke, J. M., 1995, "Ribozyme Chemistry" in *Molecular Biology and Biotechnology,* VCH Publishers, Inc., New York.

Cech, T. R., 1993, "Catalytic RNA: Structure and Mechanism," *Biochem. Soc. Trans.* 21: 229–234.

Chaplin, M.F. and Bucke, C., 1990, *Enzyme Technology,* Cambridge University Press, Cambridge.

Constabel, F., 1978, "Development of Protoplast Fusion Products Heterokaryon and Hybrid Cells," in *Frontiers of Plant Tissue Culture*, University Calgary Press, Canada.

Curtin, N.E., 1983, "Harvesting Profitable products from Plant Tissue Culture," *Biotechnology*, 1:649–657.

Dale, J. W., 1991, *Genetic Manipulations–Techniques and Applications*, Blackwell Scientific Publications.

David, S. B. and Vasaikar, S. M., 1986, *Plant Tissue Culture, Genetic Manipulation and Somatic Hybridization of Plant Cells*, BARC.

David, T. S., Anthony, J. F. G., Jeffery, H. M. and Richard, C. L., 1985, *An Introduction to Genetic Analysis*, 3rd ed., W. H. Freeman and Company, New York.

Davis, B. D., 1991, *The Genetic Revolution: Scientific Prospects and Public Perceptions*, Johns Hopkins Univ. Press, Baltimore, Maryland.

De Robertis, E. D. P. and De Robertis, E. M. F., 1988, *Cell and Molecular Biology*, L. E. C. and Febiger Publishers, U.S.A.

European Federation of Biotechnology, 1995, The Application of Human Genetic Research, Briefing Paper three, Task Group on Public Perceptions of Biotechnology.

Fedor, M. J. and Westhof, E., 2002, "Ribozymes: The first twenty years," *Mol. Cell*, 10:703–704.

Fincham, J. R. G. and Ravter, T. R., 1991, *Genetically Engineered Organisms–Risk and Benefits*, University Press, Buckingham.

Freshney, R.I., 1992, *Animal Cell Culture: A Practical Approach*, 2nd ed., Oxford Univ. Press, Oxford.

Frischauf, A.M., 1991, "Construction of Genetic Libraries in Lambda and Cosmid Vectors," in *Essential Molecular Biology–A Practical Approach*, Vol. II, IRL Press, Oxford.

Gerald C., 1996, *Cell and Molecular Biology – Concept and Experiment*, John Wiley and Sons, Inc., U.S.A.

Golding, J.W., 1993, *Monoclonal Antibody:Principles and Practice* 3rd ed. Academic Press, New York.

Hardy, K.G., 1987, *Plasmid*, IRL Press, Oxford.

Higgis, I.J., Best, D.J. and Jones, J., 1985, *Biotechnology–Principles and Applications,* Blackwell Scientific Publishers, Oxford.

Howe, C.J. and Ward, E.S., 1991, "DNA sequencing," in *Essential Mol.Biol. A Practical Approach,* Vol. II, IRL Press, Oxford.

John, E.S., 1996, *Biotechnology,* 3rd ed., Cambridge University Press, Cambridge.

Kennedy, A.C., 1996, *Molecular Biology of Biological of Control of Pest and Diseases of Plants,* CRC Press, Inc.

Kocher, T. D. and Wilson, A. C., 1991, "DNA amplification by polymerase chain reaction," in *Essential Molecular Biology–A practical approach,* Volume II, IRL Press, Oxford.

Leach, C. K. and Van Dam–Mieras, M. C. E., 1994, "Biotechnological Innovations" in *Energy and Environment Management,* Butterworth-Heinmann Ltd., Oxford.

Lewin, B., 1995, *Genes V,* Oxford University Press, New York.

Leycett, G.W. and Grieson, D., 1990, *Genetic Engineering of crop plants,* Butterworths, London.

Meyers, R.A., 1995, *Molecular Biology and Biotechnology: A comprehensive Desk Reference,* VCH Publishers, Inc., New York.

Morgan, P., 1991, *Biotechnology and Oil Spills,* Group Public Affairs, Shell Industrial Petroleum Company Ltd. Shell Centre, London.

Moss, J. P., 1992, *Biotechnology for Crop Improvement in Asia,* ICRISAT, Patancheru, India.

Mulligan, R.C., 1993, "The basic Science of gene therapy," *Science* 260: 926–932.

Old, R.W. and Primrose, S.B., 1990, "Principles of Gene manipulation," *An introduction to Genetic Engineering,* Blackwell Scientifice Publishers, London.

Pinkert, C. A., 1994, *Transgenic Animal Technology:A Laboratory Handbook,* Academic Press, San Diego, California.

Prusiner, S. B., 1982, "Novel Proteinaceous infectious Particles cause Scrapie," *Science* 216:136–144.

Rose, N.R., Milgrom, F. and Carel, J.V.O., 1979, *Principles of Immunology,* 2nd ed., Mac Millan Publishers Co., Inc., New York.

Singer, M. and Berg, P., 1991, *Genes and Genomes,* Uni. Sci. Books, Mill Valley, California and Blackwell Scientific Publishers, Oxford.

Spencer, J. H., 1991, "DNA-Protein Interaction," in *Essential Molecular Biology–A practical Approach,* Vol. II, IRL Press, Oxford.

Stanbury, P.F. and Whitaker, A., 1984, *Principles of Fermentation Technology,* Pergamon Press, London.

Steve, P., 1985, *Biotechnology–A New Industrial Revolution,* Orbic Publishers Ltd., London.

Straughan, R., 1989, The Genetic Manipulation of Plants, Animals and Microbes, The Social and Ethical Issues for consumers:A Discussion Paper, National Consumer Council, London.

Strickberger, M. W., 1976, *Genetics,* 2nd ed., Mac Millan Publishing Co. Inc. New York.

Stryer, L., 1988, *Biochemistry,* 3rd ed., Freeman, New York.

Swamy, V.M.K., 1995, Opportunities in Biotechnology and Instrumentation. Biotech. Dev. Rev.

Theil, E., 1995, "Automation in genome research" in *Mol. Biology and Biotech,* VCH Publishers, Inc. New.Y.

Vaish, N.R., Jadhav, V. R. *et al*, 2003, Zeptomole detection of viral nucleic acid using a target activated riobozyme. In RNA Society - Published by Cold Spring Harbor Laboratory Press, U.S.A.

Vasil, I.K. and Thorpe, T. A., 1994, *Plant Cell and Tissue Culture,* Kluwer Academic Publishers, London.

Watson, J.D. *et al,* 1987, *Molecular Biology of Gene,* 4th ed., The Benjamin/Cummings Publishing Company, Inc. U.S.A.

Wymer, P., 1990, Making Sense of biosensors, NCBE Newsletter, Reading Uni., Reading.